Computer Applications
in Geography

Computer Applications in Geography

Paul M. Mather
Professor of Geography
University of Nottingham
England

JOHN WILEY & SONS

Chichester · New York · Brisbane · Toronto · Singapore

Library of Congress Cataloging in Publication Data:

Mather, Paul M.
    Computer applications in geography / Paul M. Mather.
      p. cm.
    Includes bibliographical references and index.
    ISBN 0-471-92615-9 (pbk.)
    1. Geography—Data processing.   I. Title.
  G70.2.M368 1991
  910'.0285—dc20   90–43734
                              CIP

British Library Cataloguing in Publication Data:

Mather, P. M. (Paul Michael), *1944–*
  Computer applications in geography.
  1. Geography. Applications in computer systems
  I. Title
  910.285

  ISBN 0-471-92615-9

Typeset by Dobbie Typesetting Service, Tavistock, Devon
Printed in Great Britain by Courier International, Tiptree, Essex

Contents

Preface

Today's society relies heavily on computers. Economic, social, scientific, administrative and leisure activities are increasingly dependent upon computers for such things as stock control, personnel management, banking, newspaper production, airline ticketing, driving licence and social security registrations, air traffic control and office administration. Digital audio, digital telephone exchanges, fly-by-wire airliners and criminal investigation are other examples of the areas of life in which computers are now being used.

Academic disciplines are not isolated from developments taking place in the rest of society. Those who research, teach and learn in universities and colleges consciously or unconsciously modify their perceptions of the ways in which their academic work impinges upon or is, in turn, influenced by changes in society at large. New geography graduates may, perhaps, evaluate these changes in terms of employment opportunities, while teaching and research staff assess ways in which new technology can either assist them in their present work or open up new research opportunities.

Geography students are now taught to use, understand and accept computers in their everyday work. Employers of geography graduates assume that they can use computers for word processing, database management and communications, among other things. There is thus a technology push affecting the way in which geography is taught at first-degree level and an employment pull, affecting the skills and abilities expected or even required of new graduates.

That is not the only influence that the growing availability of cheaper and more powerful computers has had, and will continue to have, on geography as an academic discipline. I believe that the geographer's perception of the nature of the discipline is changing and that this change is the result of the new vistas being opened up by developments in information technology. Geography is becoming a supplier of knowledge and quantitative data concerning the human and physical characteristics of the Earth's land and ocean surfaces and atmosphere. Questions such as "where is such-and-such a feature located?" or "why is it there?" are becoming more important, while more complex relationships between phenomena are being studied at global and regional levels, as shown by current concerns with ecosystems and global climatic change. The data needed to answer those questions are being made more readily available in digital or computer-readable form, and techniques to process and analyse these data are being developed.

The concept of geography as a spatial information technology is already having a significant effect on the discipline. It is, however, important that this new view of geography should not obscure the fact that knowledge is always more important than data. Geographical research should not focus solely and exclusively on technological questions (that is, how knowledge should be applied to real-world problems) but should give considerable weight to the acquisition of new knowledge. This latter process is aided by the technological tools at our disposal, but the attraction of the tools should not obscure the fact that the aim of any academic discipline is the preservation of existing knowledge and the acquisition of new knowledge.

It is in the context of the view of geography as spatial information technology that this book is written. It is intended for students of geography at undergraduate level who wish to familiarize themselves with the terminology of computers and to read about ways in which computers are presently being used in geography. It assumes no prior knowledge of computers and no mathematical skills beyond those possessed by the average layman. The reader is warned that this is not a handbook for researchers, who should be capable of going direct to the primary literature. Specialists in particular areas of geography may find my approach over-simple. I do not apologize for this, for in my experience specialists tend to overestimate the level of knowledge and confidence possessed by the student moving from the sixth form to university or college.

The first two chapters form a technical introduction to computers and data. Chapter 1 is a review of the hardware and software components of a digital computer, using personal computers (PCs) as a model. The Microsoft MS-DOS operating system is described and examples of the use of the Wordstar word-processor and the ARC/INFO geographical information system package are given. Computers cannot operate without data, but the information that can be derived from data is dependent on the scale of measurement used to characterize the data and on the way the data are structured. These topics are discussed in Chapter 2 with special reference to spatial data.

The remaining five chapters are devoted to individual topics representing a selection of the major areas of computer use in geography. Statistical analysis has been an important component of geographical research and teaching since the early 1960s, but it is only in the last 10 years that comprehensive program packages, and the necessary computer resources, have been available to geographers, thus removing much of the drudgery associated with statistical calculations. SPSS (Statistical Package for the Social Sciences) is used in the examples in Chapter 3, in conjunction with a set of data, the World Data Matrix, measuring 14 economic and demographic characteristics of the 100 largest countries of the world in terms of population. This dataset is reproduced in Appendix A and is available on disc in DOS-compatible form.

Maps and map projections have traditionally been a central focus of geography. Chapter 4 deals with computer cartography and introduces examples using two widely available program packages – SYMAP and GIMMS. These examples use a digitized map of the countries of Africa and the World Data Matrix employed in Chapter 3. The digitized map is listed in Appendices G and H and is also available on disc. A more recent development in geography has been the use of remotely-sensed data from satellites to map the surface of the Earth. An introduction to this topic is presented in Chapter 5.

Chapter 6 provides an introduction to what is possibly the most interesting aspect of the use of computers in geography, namely, the development and use of simulation models. Examples from demography, traffic management, climatology, geomorphology and hydrology are discussed in this chapter. The final chapter attempts to integrate the subject matter of the first six chapters within the context of Geographical Information Systems. The concept of layers of information concerning places together with the data management capabilities of the computer and new methods of cartographic display have opened up new vistas for the geographer and, I believe, will reorientate the subject as a multivariate, integrating discipline focusing on the central concept of place.

The data disc mentioned in this volume is available in MS-DOS format from the author: Professor Paul M. Mather, Department of Geography, The University, Nottingham NG7 2RD.

This book is the outcome of over 20 years of teaching and research at Nottingham University. During that time I have had the pleasure of working with many interested and stimulating students and the benefit of association with amenable and lively colleagues, in particular John Cole, Roy Bradshaw, Roy Haines-Young and Michael McCullagh. I would like to express my thanks to Dr Jo Mano of the State University of New York for commenting on an early version of the manuscript and for making many worthwhile suggestions. I am grateful also to the cartographic staff of the department, Chris Lewis and Elaine Watts, for their help with the diagrams. David Ebdon gave welcome assistance with the examples in Chapter 3. None of those mentioned can be held responsible for any error that may be present in the text. Finally, I must thank my family for tolerating my long absences. I will leave the reader to judge whether the effort was worthwhile.

PAUL M. MATHER
Nottingham
January 1990

CHAPTER 1

Introduction to Computers

1.1 INTRODUCTION

A computer is an electronic device that stores and manipulates numerical or textual data in a way that is specified by instructions called programs. Computers are used to assist geographers in carrying out a variety of tasks, provided appropriate programs and data are available. Some examples of operations that computers can perform are:

1. Find the mean population of all the states of the USA on a certain date, given the population figures for each county.
2. Retrieve from a list of countries of the world all those that have a gross national product per head of less than US$800 and a life expectancy at birth of less than 50 years.
3. Calculate the correlation between rent charged for housing and distance from the city centre for a given city.

The computer-assisted solution of each of these problems involves the following steps:

- Input of data to the computer, for example, the US population by county, details of the socioeconomic and demographic characteristics of the countries of the world, or location and characteristics of rented accommodation in a city.
- Manipulation of data, either arithmetically or logically. Numerical manipulation involves calculations such as finding the mean of a set of data, while logical manipulation involves comparisons of the type "is A equal to B?".
- Output of results. In the three examples above, output would take the form of a printout or a display on a terminal screen.

These operations are described in more detail below.

There are other examples of the way in which computers are used in geography. A computer can control the movements of a pen (or a number of

1

pens) on a plotting device, which is described in Chapter 4. Given the relevant data and programs, the computer can

4. Draw a contour map of the mean annual rainfall isohyets for the Continental USA given mean rainfall values and map coordinates of a number of rainfall gauges spread across the USA.

A satellite picture such as the one used on television weather forecasts is made up of an array of numbers called a digital image. Digital images obtained from satellites are discussed in Chapter 5. A computer can control a device which converts these numbers into a form suitable for input to a colour TV monitor. The digital image can be transformed in various ways to improve its visual appearance. Another operation that might be carried out using a computer is:

5. Display a colour satellite image of London and manipulate the colours in the image in such a way as to enhance the appearance of the image for visual interpretation.

If the behaviour of a system such as a river basin or a city transport network can be defined in sufficient detail, then aspects of its behaviour can be represented by a digital simulation model (Chapter 6). If the structure of the system, and the processes controlling its behaviour are known, then it becomes possible to:

6. Use a computer model of the system to simulate its operation, allowing the user to perform experiments by changing conditions or by altering the inputs to the system.

The computer can, therefore, act as a calculator, a filing system, a map production system, an image display and enhancement system and a geographical laboratory. Although the truth of this statement has been evident for some years, the size and cost of computers in the 1960s and 1970s was such that they were not readily available for everyday use. Until the introduction of microcomputers (personal computers) in the 1980s, and the demonstration by schoolchildren that using a computer was not an impossible feat, many geographers did not consider using a computer to assist them in their work because of cost and accessibility problems. Big ("mainframe") computers of the 1960s such as the IBM 360 (Figure 1.1) were kept in central computer centres and tended by operators. Access to the mainframe computer was rationed and, in many universities and colleges, potential users from departments other than mathematics, science and engineering were not actively encouraged. Nowadays, most universities and colleges still keep a central computer, which is often a large mainframe, for "number-crunching" operations and large-scale database manipulation, but many geography departments have invested in personal computers for use in teaching and research (Figure 1.2). One advantage of the personal computer is that it makes access to computing power relatively easy;

Figure 1.1 IBM Series 360 Model 22 computer. Machines such as this were widely used during the late 1960s and early 1970s. They were normally housed in air-conditioned rooms, tended by specialist operators and rarely approached by users. Hollingdale and Toothill (1965) give a readable account of computing during its early stages of development. (Photograph courtesy of IBM UK Ltd.)

furthermore, personal computers can be linked to the central mainframe by a local-area network (Section 1.8.2) so that users get the best of both worlds.

In order to use a computer sensibly it is not necessary to be able to program the computer, or even to understand how it works at the physical or electronic level. It is, however, important that the user has some appreciation of the way in which a computer operates at a logical or conceptual level, and it is the purpose of this chapter to provide the necessary background.

1.2 DATA INPUT

Before data can be processed by a computer they must be fed in, or input, to the computer system. Data can be input from a terminal by pressing the appropriate keys. This is the standard way of supplying small quantities of numerical or textual data.

Map data (made up of the xy coordinates of points and lines, together with values of characteristics such as height above sea-level associated with these points) are input to the computer by a procedure called *digitizing*. This process

Figure 1.2   IBM PS/2 Model 30 personal computer with colour display and keyboard. The fixed disc is contained within the processor box, while the floppy disc drive is located in the upper section of the processor box on which the display screen stands. (Photograph courtesy of IBM UK Ltd.)

is described in more detail in Chapter 4; one example is given here and illustrated in Figure 1.3. The map to be digitized, or converted to computer-readable form, is placed on a table called a digitizing table. The position of a pointer can be sensed by the computer so the xy coordinates of points of interest can be collected by using the pointer while values or codes associated with these points are provided manually using a keyboard. Other methods of digitizing include automatic scanning procedures. Black-and-white aerial photographs can be turned into digital pictures by the use of such scanning methods, which detect the shade of grey at each of a large number of regularly-spaced points on the photograph and represent the level of grey at each point by a number on a scale ranging from 0 (black) to 255 (white). The analogue photograph is thus turned into a digital image, that is, an array or table of numbers in which the value at any point in the array corresponds to the level of grey at the corresponding point in the photograph.

Some data sets may already exist in a form suitable for direct input to the computer. Examples are:

Figure 1.3   Map digitizing in progress. The operator is inspecting a digital map display on the screen of a Digital Equipment Corp. VAXstation computer (the processor box is just visible below the desk). The map being digitized is lying on a digitizing table, and map coordinates (defining the positions of map features) are collected by placing the pointer (in this case formed by black cross-wires on a transparent plastic disc) over the point of interest and pressing one of the buttons on the keypad attached to the pointer. The coordinates of the point are sent along the wire joining the pointer/keypad to the computer and the screen display is updated. (Photograph courtesy of LaserScan Laboratories Ltd, Cambridge.)

- The UK Census of Population from 1981, which is available in computer-readable form for enumeration districts (census tracts) and for 1 km grid squares.
- Remotely-sensed images from satellites such as Landsat (described in Chapter 5) are relayed back to Earth in numerical form and are stored on magnetic tapes suitable for reading directly into a computer.
- The results of experiments in the laboratory or in the field can be recorded on a computer-readable medium using a device called a data logger. The data logger is, in fact, a portable microcomputer which can be instructed to interrogate suitably-interfaced instruments at a set time interval, and record the readings. The recording can be collected at a convenient time and fed into the computer for analysis.

An example of the kind of data capture described above is the use of a data logger to interrogate meteorological instruments at 30-minute intervals and record measurements such as barometric pressure, temperature and rainfall.

1.3   DATA REPRESENTATION

1.3.1   Base two, base 10 and base 16 numbers

Data that are input to a computer are stored and manipulated by the computer using a coding system. Written language also uses a coding system; this book is written in one such system which is called the Latin alphabet (though Arabic numerals are used rather than the I, V, X, L and C of the Romans). Computers are electronic devices which recognize only two states, corresponding to on and off or 1 and 0. Consequently, a computer cannot use the Latin alphabet (which has 26 characters) or the Arabic number system (which has 10 digits). The data that a computer stores and manipulates are represented by a coding system that has only two characters – 0 and 1. It is called the *binary code*.

The idea of a two-symbol code is not new. Morse code uses the dot ($\cdot$) and dash (-) to convey information; for example, $\cdot \cdot \cdot - - - \cdot \cdot \cdot$ means "SOS", the international distress call. The restriction to two symbols makes arithmetic rather tedious for people to follow; we use 10 digit symbols (0, 1, 2, . . ., 9) and base our arithmetic on the decimal number system while computers use two symbols (0 and 1) and base their arithmetic on the binary number system. Computers contain instructions which allow them to translate from their internal binary code to the decimal code that people use.

The operation of the binary code will be explained using an example. Consider the number 352 (there is nothing special about 352; it was selected arbitrarily). The "3" really means three hundreds. The "5" means five tens, and the "2" means two ones or units. The number 1 can be expressed as ten to the power of zero (10^0). Using the same notation the number 10 can be written as 10^1

while the number 100 can be expressed as 10^2. Thus, the number 352 could be written in full as $(3 \times 10^2) + (5 \times 10^1) + (2 \times 10^0)$. The position of each digit in an ordinary decimal number implies a multiplication factor which is a power of ten. We do not need to be told explicitly that the 3 in 352 means 3 hundreds (or 3×10^2). If the number is written out in full, though, it is clear that (a) 10 is the base of our number system and (b) the position of each digit in a decimal number implies multiplication by a power of ten. The first digit to the left of the decimal point (explicit or implied) is multiplied by 10^0, the second by 10^1, the third by 10^2 and so on. Such a method of representation of numbers is called positional, and differs from the Roman system in which (for example) in the number MDCCCXII each of the three C's is equal to 100 regardless of its position.

The binary system uses only two symbols, 0 and 1. The binary and decimal systems differ only in that the base of the binary system is two rather than ten, and the position of a digit implies multiplication by a power of two. Thus, the number 101 in the binary system really means $(1 \times 2^2) + (0 \times 2^1) + (1 \times 2^0)$. Thus, 101 in binary notation is equivalent to the decimal number 5. To avoid confusion a subscript 2 or 10 should be appended to every number to indicate the base of the system in which that number is expressed (for example 352_{10} or 101_2). It should be appreciated that other number systems are also possible (e.g. base 8 or base 16) and are indeed used for some purposes. Base 16 or *hexadecimal* numbers are described next.

A base two number system is difficult for humans to handle efficiently. A large decimal number – 65535_{10}, for example, translates to 1111111111111111_2, which is a little cumbersome. To make for easier communication between people and computers an apparent complexity is introduced in the form of yet another number system which, fortunately, is as simple as the base two or base ten system. The base of this new system is 16 and it has 16 digits – 0 to 9 followed by A to F. Table 1.1 shows the base ten numbers 0–16 expressed in base two and base 16 form. You can use this table to translate any base two number to its base 16 equivalent. Simply split the base two number into groups of four digits, starting from the right-hand end and adding zeros to the left-hand end if necessary. The base two number 1000111101111110 is split up into four groups of four digits as follows: 1000 1111 0111 1110. Next the individual sets of four digits are translated into base 16 digits using Table 1.1. This gives 8F7E, which is much easier to write down and remember than its base two equivalent. The name given to the base 16 number system is *hexadecimal*, often shortened to "hex". You will come across this term frequently if you read any computer magazines.

Modern microcomputers such as the Apple Mackintosh and the IBM PS2 are able to store a lot of numbers. These numbers are held in the computer's *random-access memory* (RAM), the capacity of which is described in terms of the number of *bytes* that it can hold. A byte is a set of eight binary digits

Table 1.1   Different ways of expressing numbers: the decimal (base 10) numbers are shown in the centre column with their base 2 equivalents on the left and the base 16 equivalents on the right

Base of number system			Base of number system		
2	10	16	2	10	16
0	0	0	1000	8	8
1	1	1	1001	9	9
10	2	2	1010	10	A
11	3	3	1011	11	B
100	4	4	1100	12	C
101	5	5	1101	13	D
110	6	6	1110	14	E
111	7	7	1111	15	F
			10 000	16	10

(0s or 1s); each byte is capable of storing a number ranging in magnitude from 00000000_2 to 11111111_2 or 00_{16} to FF_{16}. Negative numbers can also be stored, but they are still represented by 0s and 1s. Usually the range of numbers stored in a single byte is such that both positive and negative numbers are included. The positive numbers 0–127 are represented by the bit patterns 00000000 to 01111111 and the negative numbers -128 to -1 by the patterns 10000000 to 11111111. A similar system is employed for numbers held in multiple byte groups; thus, a two-byte storage unit holds numbers in the range 0 to 32767 and -32768 to -1. Descriptions of computers generally contain some reference to the size of the memory, for instance 32 Kb, 512 Kb or 1 Mb. The b stands for byte while the K stands for kilo, which in this context means 2^{10} or 1024. The symbol M stands for mega – one megabyte (Mb) is 1024 Kb.

1.3.2   Computer memory organization

The random-access memory of the computer can be used to store small numbers in single bytes, while larger numbers are held in groups of contiguous bytes. Numbers with fractional parts (such as 3.62) are also held in a group of bytes, normally four. Individual bytes can also hold instructions, which tell the computer what to do, or they can hold character data. The statement "I like ice cream" would be stored in 16 bytes. This statement is, of course, to be stored in the computer for later retrieval. It is not an instruction for the computer to obey. It is important to remember that numerical and character data as well as instructions are all stored in the computer's memory as sets of binary digits (0s and 1s). The computer does not know whether a particular byte holds a small number, part of a big number, a character or an instruction. Such information is provided by the controlling program (Section 1.4). Figure 1.4 shows a typical memory layout in schematic form.

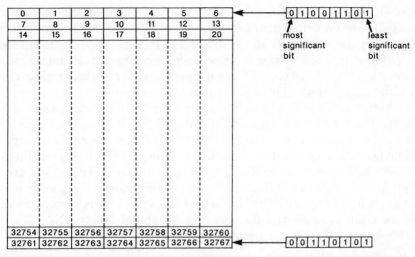

Figure 1.4   Schematic layout of a 32 Kb random-access memory. Individual byte addresses run from 0 to 32 767. The contents of each byte consist of eight binary digits (bits); the bit-pattern may represent a number or an ASCII character (see text for explanation). A distinction is made between the address of a byte in memory and the information stored in that byte. In the diagram the byte whose address is 6 contains the base 2 number 01001101, while the byte at address 32 767 holds the base 2 number 00110101.

1.3.3   Storage of character data – the ASCII code

As we have already seen, characters are stored in the memory of the computer in numerical form. A code called the American Standard Code for Information Interchange (usually shortened to *ASCII*) is frequently used. In the ASCII code the letter A is stored as 65_{10}, B is 66_{10} and so on to the letter Z, which is represented as 90_{10}. Altogether there are 128 ASCII codes which provide for all the keys on the keyboard of a computer terminal plus some special non-printing characters (such as "delete last character" (code 127_{10}) or "ring terminal bell" (code 7_{10})). Each ASCII code occupies one byte. Character (text) data are converted into ASCII form on input and are converted back to readable form (characters, digits and punctuation marks) for printing or for display on a screen.

1.4   INSTRUCTING THE COMPUTER

1.4.1   Machine code and assembler languages

Instructions are held in the computer's memory. When an instruction is required it is brought from the memory to a special part of the computer called the central processing unit or CPU. The CPU performs calculations and other manipulations on data stored in memory. Facilities for these operations are provided by electronic units which can add, subtract, multiply, divide and make comparisons between numbers. These operations are carried out in special memory locations called registers. Many computers have a special register called an accumulator which is used for arithmetic operations.

A short example will help to explain how the CPU operates. Assume that two numbers are stored in memory cells labelled 70_{10} and 71_{10} (see Figure 1.4). We want to add these two numbers together and store the result in memory cell number 72_{10}. To simplify matters, it is assumed that the sum of the two numbers held in memory cells 70_{10} and 71_{10} is non-negative and will not be larger in magnitude than 127_{10}, the largest positive number that can be stored in a single byte (memory cell). This assumption will allow us to store the data and the result in single-byte memory cells, numbered 70_{10} to 72_{10}.

The instructions that are given to the CPU will be:

1. Bring the number stored in memory cell 70_{10} into the accumulator.
2. Bring the number stored in memory cell 71_{10} and add it to the number already in the accumulator.
3. Put the result into memory cell number 72_{10}.

Notice that we distinguish carefully between the position of the memory cell (70_{10}, 71_{10}, 72_{10}) and the number stored in that cell. The position of a memory cell is called its address while the number stored in the cell is the contents of that address.

The set of instructions labelled 1, 2 and 3 above is an example of a computer program, which is a sequence of commands designed to cause a CPU to carry out a particular operation, in this case addition of the contents of two specified memory cells. A computer program is not written in the form of English text statements as shown in the example above. A special language, called a *programming language*, is used. Each different make of CPU has its own instruction language (or *machine code*). The MOS Technology 6502 CPU is widely used in small microcomputers; an example of its machine code will be given shortly. Other well-known CPUs are Zilog's Z80 and Intel's 8086, which is used in the IBM PC. The larger versions of the IBM Personal System/2, which replaced the PC in 1987, use Intel's 80386 or 80486 CPU while the Apple Mackintosh is built around a Motorola 68000 CPU. The different makes of CPU (such as Motorola, Zilog and Intel) have different instruction sets, as do the various processors produced by each individual manufacturer. Thus, the instruction set of the Intel 8086 is not the same as that used by the Intel 80386. This point is considered in more detail at a later stage.

The 6502 instruction set, like other machine codes, associates particular operations with specified numerical codes. Thus, "bring a number stored in a specified memory cell and put it in the accumulator" has the code 10100101_2

or $A5_{16}$. The code for "bring a number from a specified memory cell and add it to the number already stored in the accumulator" is 01100101_2 or 65_{16}, while "take the number stored in the accumulator and put it in a specified memory cell" is 10000101_2 (85_{16}). In each case the address of the specified memory cell is placed after the instruction code. A 6502 CPU would therefore interpret the binary codes

$$10100101_2 \qquad 01000110_2$$
$$01100101_2 \qquad 01000111_2$$
$$10000101_2 \qquad 01001000_2$$

to mean "place the contents of memory cell 70_{10} (01000110_2) in the accumulator. Then add to the accumulator the contents of memory cell number 01000111_2 (71_{10}) and, finally, put the result in memory cell number 01001000_2 (72_{10})". These instructions and addresses are a little less cumbersome when expressed in hexadecimal notation:

$$A5_{16} \qquad 46_{16}$$
$$65_{16} \qquad 47_{16}$$
$$85_{16} \qquad 48_{16}$$

Nevertheless, it would be a considerable task to remember all the numerical instruction codes that are available, so writing a program of any length would be a task not to be embarked upon lightly.

To make things slightly easier, each instruction code is associated with an abbreviation or mnemonic which is relatively easy to remember. These mnemonics are collectively termed the *assembler language* of the computer. In 6502 assembler our little program could be written as

$$\text{LDA} \qquad \# 70_{10}$$
$$\text{ADC} \qquad \# 71_{10}$$
$$\text{STA} \qquad \# 72_{10}$$

which is certainly less formidable in appearance than the machine code. The mnemonic LDA stands for load accumulator (place in the accumulator the contents of the specified address, in this case byte 70), ADC means add to the contents of the accumulator the number stored in the specified address, and STA means store the number held in the accumulator in the given address. Even though assembler language programs are easier to write and to comprehend than the equivalent machine-code program, the use of assembler language is tedious and time-consuming, so many computer users would be put off if they had to write programs in an assembler language. Furthermore, since each type of CPU has its own instruction set, a given program would have to be rewritten if it were to be used on a CPU other than the one for which it was developed.

1.4.2 FORTRAN and BASIC

When computers were still in their infancy, a group of scientists at IBM realized that programming in machine code or assembler language was restricting the number of people who could use computers. They decided to develop a computer language that was relatively easy for scientists to use. This language became known as FORTRAN. The word stands for FORmula TRANslation. It is still the most widely-used computer language for scientific and technical applications, and has been adopted by manufacturers of computer systems all over the world. A few years after the introduction of FORTRAN a group of researchers at Dartmouth College in the United States developed a language that was simpler to use than FORTRAN. They called it BASIC (for "Beginner's All-purpose Symbolic Instruction Code") and it is now widely used in schools, colleges and universities as well as in business applications. Its attraction can best be demonstrated by giving an example of a BASIC program that will find the area of a circle using the formula area $= \pi \times r^2$, an operation which is rather more complicated than the earlier example of adding two numbers together.

```
10 PRINT "What is the radius of the circle?"
20 INPUT R
30 A = PI * R * R
40 PRINT "Its area is", A
50 END
```

Most readers will be able to work out how this program operates without assistance. PRINT causes the text enclosed in quotation marks ('') to be output to the user's terminal. INPUT reads the number that the user types at the keyboard, while END signals the end of the program. The symbol * means "multiply". If this program were to be entered into a computer and then run, the result would appear as follows:

```
What is the radius of the circle?
?6.24
Its area is 122.3261
```

The ? symbol means that the computer is waiting for input to be typed at the keyboard.

A computer, however, cannot directly execute instructions written in BASIC or any other language; it can only execute instructions expressed in binary machine code. A BASIC or FORTRAN program must therefore be translated into binary machine code before the CPU can execute it. This translation process is carried out by a special program called a *compiler* (in the case of FORTRAN) or an *interpreter* (in the case of BASIC). A compiler converts the whole of the FORTRAN program to the machine code of the CPU being used in one go, while the interpreter translates each line into machine code as required. Computer

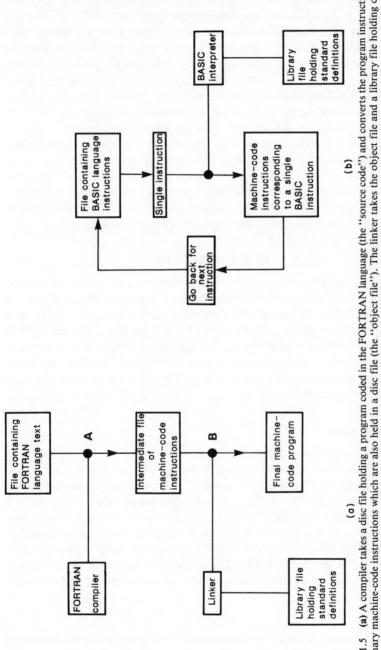

Figure 1.5 **(a)** A compiler takes a disc file holding a program coded in the FORTRAN language (the "source code") and converts the program instructions into binary machine-code instructions which are also held in a disc file (the "object file"). The linker takes the object file and a library file holding code for standard definitions (such as square root, logarithm and cosine). Those functions used in the program are extracted from the library and added to the object file to produce a machine-code program that the CPU can execute. A compiled program runs faster than the equivalent interpreted program (Figure 1.5(b)) but if errors are found then the FORTRAN source code must be amended and the compilation/linking process repeated. Languages other than FORTRAN can be compiled; examples are Pascal, C, Algol and some versions of BASIC. **(b)** Each instruction in a BASIC program is converted into a set of machine-code instructions. Any standard definitions (such as square root, logarithm or cosine) that are required are brought from a library file. The machine-code instructions are executed and then lost as the next BASIC instruction is interpreted. Errors can be detected and corrected quickly but large programs run more slowly than their compiled equivalents.

manufacturers supply FORTRAN compilers and BASIC interpreters to convert programs written in these languages into the machine code used by their particular CPU. Figure 1.5 illustrates the processes of compilation and interpretation.

The first advantage of high-level languages such as FORTRAN and BASIC (and others such as Pascal, Cobol, Algol and C) is that they are much easier to use than the assembler languages described earlier. The second advantage is that programs written in standard FORTRAN or BASIC will run on any computer having a FORTRAN compiler or a BASIC interpreter, whereas a program written in 8086 assembler will only run on an Intel 8086-based computer. FORTRAN and BASIC programs are *portable* (i.e. they can be moved from one computer to another and still work). Another word for portable is *machine-independent*. Given the high costs involved in developing large programs for operations such as statistical analysis, mapping, or processing of remotely-sensed data, the advantages of writing in a portable language are overwhelming. Assembler languages are only used for specific applications where either execution speed or the ability to perform some non-standard operation is important.

1.5   DISCS AND TAPES

Newspapers occasionally carry stories to the effect that the police records of a certain city, or all the maps produced by the US Geological Survey, are stored on a computer. Clearly this is not possible if the random-access memory of even a large computer is capable of holding only a few million bytes. The computer's random-access memory is reserved for storage of data that are in immediate use, and for storage of programs that are currently being executed. All other data and programs are kept on external storage in the form of floppy or hard discs and magnetic tapes.

1.5.1   Floppy discs

The most common form of external storage for microcomputer systems such as the IBM PC and the Apple Macintosh is the *floppy disc*. A disc is a circular plastic sheet coated with a magnetic material. The disc surface can be in one of two states of magnetism at any point and this property is used to store a 1 or a 0 at each of a large number of points on the disc surface. Recall that a byte is a group of eight binary digits or bits, so the points are grouped in sets of eight. This means that there is a direct correspondence between a byte in a memory cell and a byte on disc. The contents of each byte can be moved back and forth from disc to memory along a connector called a data bus.

1.5.2   Files

A related set of bytes, such as the Census data for a particular city, is held on the disc in a logical structure called a *file*. A file is really like a book on a shelf.

It is a related set of information, which could be numerical or character data or a set of program instructions. For example, the characters forming this chapter are held in a file. Each file is given a unique name so that it can be located when needed. A single disc can hold a number of files, depending on the capacity of the disc (measured in bytes) and the characteristics of the operating system (which are considered shortly). A floppy disc normally has a storage capacity of between 100 Kb and 1.4 Mb. The floppy disc can be removed from its drive and another disc put in its place, so the effective capacity of the disc system to store data and programs is very considerable. The standard IBM PC uses floppy discs with a 5.25-inch diameter; each disc can hold 360 Kb. The IBM Personal System/2, introduced in 1987, uses 3.5-inch floppy discs with a storage capacity of 1.4 Mb (Figure 1.2).

1.5.3 Hard discs

Bigger computers generally process more information and therefore need access to more external data storage than does a typical office or school computer. Large-capacity discs are used with these bigger computers, and the capacity of these discs is measured in megabytes. Whilst floppy discs are very tolerant in terms of their environment, big discs are not. The disc is read and written by a magnetic head that flies very close to the disc surface which is rotating at a speed normally of 3000 revolutions per minute. A piece of dust from tobacco smoke, a human hair or a water droplet might get between the read/write head and the disc surface, causing at best a failure to read or write data to or from the disc and at worst producing a "head crash" which will scratch the disc, damage the heads, and in general cause problems for the computer users. Hence, big discs are enclosed in sealed cases and the pressure of air inside the case is kept slightly higher than atmospheric pressure so that dust is unable to enter. This kind of disc is called a fixed disc, a Winchester disc or a hard disc. Unlike the floppy disc it cannot be replaced since it is fixed in its drive. Not all big discs are fixed discs; exchangeable big discs are widely used, but the fixed disc is most popular on small (micro) and medium-sized (mini) computers. Fixed discs of 20 Mb capacity and greater are now available for IBM PC-type microcomputers and have replaced floppy discs in many routine office applications. Besides having a larger storage capacity than floppy discs, the fixed disc is also capable of reading and writing data and transferring it to the CPU at a more rapid rate than a floppy disc.

1.5.4 Magnetic tape systems

Data stored on fixed discs are vulnerable because the disc head mechanism may crash and damage the disc surface, making some sections of the disc unusable. A strong magnetic field may also erase valuable data from the disc. Some means of archiving (preserving) data that are stored on fixed discs is thus desirable. The most widespread archiving medium is the magnetic tape. A single standard 2400-foot reel of magnetic tape can store up to 35 Mb of data and so the entire contents of a standard 20 Mb fixed disc can be stored on a single reel of tape. Minicomputers, such as the Digital Equipment Corporation VAX, often have fixed discs with capacities measured in gigabytes (1 Gb = 1024 megabytes) and the archiving problem is correspondingly greater. Magnetic tapes are examples of serial storage devices; the data stored on them are written in byte-by-byte form (Figure 1.6). Each byte is represented by a set of 0s and 1s written on the tape perpendicular to the direction of movement. Ten years ago the standard packing density was 556 or 800 bytes per inch (b.p.i.). Nowadays, the standard packing density is 1600 b.p.i. but this is being replaced by a 6250 b.p.i. standard. A 2400-foot reel of tape could, in theory, store 180 Mb of data at the 6250 b.p.i. density. In practice, however, an "inter-record gap" of 1–2 cm separates each record, which is equivalent to a line on a printed page. The actual capacity of a reel of 6520 b.p.i. tape is thus nearer 80 Mb.

As well as providing a means of archiving data held on a fixed disc, magnetic tapes provide a means of transferring data from one computer to another. If small amounts of data are involved a floppy disc can be used to transfer data,

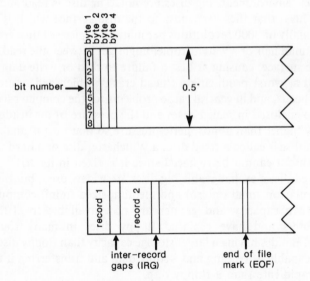

Figure 1.6   Storage of data on magnetic tape. (Top) Each byte is written across the width of the tape, at right-angles to the direction of tape movement. The ninth bit is used for error checking purposes. (Bottom) A related set of bytes, such as a line of text, form a record. A file consists of groups of records. For example, each line of this book is held as a record on a magnetic tape. The entire book is a single file. Pairs of records are separated by blank Inter-Record Gaps (IRG). Files are separated by End-of-File (EOF) marks. The last file on a tape is followed by two successive EOF marks.

providing that both computer systems (the donor and the recipient) are compatible. There are a variety of ways in which data can be written to a floppy disc; much depends on the operating system (described in the next section) and the packing density and storage capacity of the floppy disc. The IBM PC uses Microsoft's MS-DOS operating system and stores 360 Kb per disc, using both sides of the disc. The ICL Quattro microcomputer uses the Concurrent CPM/86 operating system and stores 720 Kb of data per disc. Floppy discs written on an ICL Quattro therefore cannot be read by an IBM PC. The same problems of format and packing density complicate the transfer of data from one computer to another using magnetic tape. As noted above, packing density is normally one of 550, 800, 1600 or 6250 b.p.i. while the format (the way in which data are organized on the tape) depends on the operating system. A standard format, ANSI (American National Standards Institute), is normally supported by the more widely-used operating systems and ANSI-standard tapes can normally be read by computers equipped with tape drives that can handle the particular packing density used to write the tape.

1.6   OPERATING SYSTEMS

If you have access to a computer you will know that you can type a command such as DIR to get a list of the files held on a disc, or DEL to delete a file. These commands are not part of the instruction set of the CPU. They are *operating system* commands. All except the most simple computers have an operating system which is the master program that controls the operation of the computer. When the computer is switched on, a program stored permanently in a special kind of protected memory, called read-only memory (ROM), is activated. This program brings from the disc into the main random-access memory the programs that form the operating system and so makes them available to the user. The name given to the process of loading the operating system at start-up is *bootstrapping* for the computer is, by analogy, "hauling itself up by its own bootstraps". The operating system is kept in memory while the computer is switched on and it provides the user with the facilities required to enter data, run programs, transfer data to and from disc, and to manage files on the disc. At one time all computer manufacturers produced their own operating system software; examples are the RSX-11 and VMS operating systems developed by Digital Equipment Corporation for its PDP and VAX range of minicomputers. With the coming of mass-produced microprocessor-based computers, individual manufacturers did not have the resources to develop and maintain their own operating systems so independent companies began to market operating systems such as UNIX, MS-DOS and CP/M. The IBM PC and a host of look-alikes or compatibles have standardized on the Microsoft MS-DOS operating system (true IBM machines use a version of MS-DOS called PC-DOS).

1.6.1   Introduction to MS-DOS

MS-DOS (MicroSoft Disc Operating System) is designed for microcomputers having a minimum of 256 Kb memory and one floppy disc drive. The programs making up MS-DOS are supplied on a floppy disc which is known as the system disc. The following brief description of MS-DOS assumes that a computer with one floppy disc drive and a hard disc is available.

MS-DOS consists of a set of files. The programs making up software packages, as well as the datasets on which these programs operate, are also stored in files. The full description of an MS-DOS file is given by:

- The identifier of the disc drive containing the disc which holds the file to be accessed,
- the name of the directory on that disc containing the file, and
- the name of the file.

On the system described here there are two disc drives – the floppy drive, which will be identified by the code B: and the hard disc, which is given the identifier C:. The first part of a file description is thus C: or B: depending on whether the file is stored on the hard disc (C:) or a floppy disc (B:).

The second part of the file description is the name of the *directory* containing the file. The idea of the directory is shown graphically in Figure 1.7. You might like to think of a directory as a name for a group of files, for example all

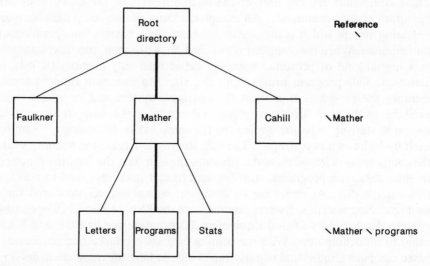

Figure 1.7   Hierarchical file structure used by later versions of MS-DOS. The master or root directory contains pointers to directories lower down the file structure. These directories may contain pointers to sub-directories and so on. For example, the reference \ *mather* \ *programs* \ *contour.for* indicates that the file *contour.for* is held in subdirectory *programs* of directory *mather* which itself links into the root directory. Other operating systems such as Unix and VAX/VMS use a hierarchical file structure.

files belonging to a user called Mildred. Mildred herself may have several different types of files (containing, for example, letters, memos and addresses), and she might like to keep these different files in separate subdirectories, called LETTERS, MEMOS and ADDRESS (a directory name must not be longer than eight characters, the first of which must be a letter).

The third part of the file description is the name of the individual file. An MS-DOS filename consists of (i) an eight-character or less filename (ii) a decimal point or full-stop and (iii) a three-character filename extension. The filename should describe the contents of the file while the extension describes the nature of the file. Some examples are:

shortroo.bas	a BASIC program (.bas) to find shortest routes in a network.
william1.let	a letter to William.
william2.let	another letter to William.
contour.for	a FORTRAN program to draw contours.

Remember that these filenames are only the third part of the complete file description which needs the disc drive identifier (C: or B:) and the directory specifications to be complete. Some examples are:

b: \ mather \ programs \ contour.for

> a FORTRAN program called contour.for, stored in directory mather and in subdirectory programs on a floppy disc.

c: \ mildred \ letters \ william1.let

> a file called william1.let in directory mildred, subdirectory letters on the hard disc.

Notice that the various links in the path (Figure 1.7) defining the full file description are separated by the " \ " symbol. The *root directory* is the master directory at the base of the tree structure (Figure 1.7). You can move downwards in the tree by using the command cd. For instance, when you switch on the machine you are logged in to the root directory. To change to directory mather type the command *cd mather* followed by ⟨return⟩. To go back to the root directory, type cd .. ⟨return⟩. The command *cd* means "change directory". It is an MS-DOS system command. All operating systems are seen by the user as sets of built-in commands such as *cd*.

Examples of other MS-DOS system commands are:

print ⟨filename⟩	prints the contents of the file specified by ⟨filename⟩
** example: print b: \ mather \ letters \ publish.let	

dir	puts listing of files stored in your current directory to the screen.
**example: dir b:	

del ⟨filename⟩	deletes specified file from the disc.
**example: del b: \ mather \ letters \ publish.let	

copy ⟨filename1⟩ ⟨filename2⟩	copies file 1 to file 2
**example: copy c: \ mather \ memos \ car.mem c: \ bradshaw \ memos \ car.mem	

Further examples can be found in the Microsoft MS-DOS *Users' Reference Manual*, which is supplied with all MS-DOS microcomputers.

1.6.2 Concurrent CPM

Other operating systems, such as Concurrent CPM/86 (CCPM) use similar commands, but the file structure is different. CCPM, for instance, divides the hard disc into 16 areas called user areas. All user areas are at the same level, so the concept of the hierarchical file structure, as used by MS-DOS, is not present. The operating system commands in CCPM are less numerous than those available in MS-DOS, and the format is sometimes different. Some command names differ from their MS-DOS equivalents; to erase a file on a CCPM system, the command is "era" (for erase) rather than "del" (for delete) on an MS-DOS system. The difference in the way files are stored on disc in CCPM and MS-DOS means that you cannot write files to a floppy disc on computer A (running MS-DOS) and read them on computer B (running CCPM). This can be a significant problem when transferring programs and data from one user to another.

1.6.3 Other operating systems

So far we have looked at desktop micros, which are "single-user" machines. In other words, only one person can use them at any one time. The exception to this among the computers described so far is the ICL Quattro which was mentioned above; it uses the Concurrent CPM/86 operating system. "Concurrent" means that the computer can run several tasks apparently simultaneously. The Quattro, in fact, can in theory accommodate four users (at four separate terminals) each running up to four tasks. Minicomputers such as the DEC VAX and mainframes such as the ICL 3900 series can handle many users at one time. Each user thinks that he/she is the only user, because the computer handles his/her commands much in the same way that an IBM PC or an Apple Mackintosh would. Yet a mini or mainframe computer is more than capable of handling the data typed in at a terminal. Most people find it difficult to type more than 300 characters per minute. This occupies only about 0.1% of the CPU's available power, so the operating systems used by mini and mainframe computers incorporate a method of sharing CPU time between multiple users. Unless the number of users approaches the maximum capability of the machine then the response time at each terminal will be at least as good as that achieved on a desktop micro. Indeed, the response time may well be significantly better, for the discs used by mini and mainframe computers can generally access data much faster than can the fixed discs used in micros, and they are certainly many times faster than floppy discs.

Because of the inherently greater complexity involved in handling simultaneous users, the operating systems used by mini and mainframe computers are rather more complicated than MS-DOS or CCPM, which are described above. Operating systems such as VMS (Virtual Memory System) as used by the DEC VAX range of minicomputers allow multiprocessing or time-sharing, so that each of a large number of users can be simultaneously managed. These users are termed *interactive* as they are running jobs (such as word-processing, described below) from their own terminal and communicating with the computer in real time. Because there are many users of such machines it is necessary to ensure that each user's files are protected so that other users cannot read, modify or delete them. It is also necessary to manage the resources of the computer system so that each user gets a fair share of computer time and disc space. The operating system looks after these tasks. Each interactive user must "log in" to the system before he/she can use any system resources. To log in the user types a username (for example MATHER, which is my username on the Nottingham University Remote Sensing VAX computer). The operating system then asks for a password. This is like the Personal Identification Number (PIN) that a bank cash card uses. If you get it wrong you are allowed a few tries before the operating system disconnects you. If the computer you are trying to access is secure (for example, a bank computer or a university administrative computer holding exam marks and personal details of students and staff) an alarm will be activated so that the system manager is aware that a "hacker" may be on the loose.

Once the user has logged in then he/she interacts with the computer in much the same way as he/she would on a micro system. Operating commands such as "dir" and "del" are available, and (on most university central systems) a wide range of applications software packages are accessible. If the job that the user wishes to run is large, either in terms of computing time or memory requirements, then it can be submitted to a *batch queue*. A job placed in a batch queue will wait its turn to run, and will, when its turn comes, run in the computer together with interactive jobs such as word-processing that are connected to human users. However, a batch job will have a lower priority than these interactive jobs and will generally use the spare capacity of the machine rather than compete for prime time. A program called the scheduler is part of multiprocessing operating systems such as VMS; its task is to ensure that each job receives the resources it needs in the fairest way.

1.7   THE COMPLETE COMPUTER

The complete computer is made up of the components that have already been described. The central unit of the hardware (that is, the physical units which make up the computer) is the CPU which interprets instructions drawn from its instruction set. These instructions normally provide for arithmetic (add, subtract, multiply and divide) operations, and for comparisons (equal, greater, less than, and so on). The CPU can access data that resides in the random-access memory. Programs (or software) are stored in the random access memory alongside data. Additional memory storage is provided by external storage units such as fixed and floppy discs. The operating system, which controls the operation of the computer, is normally read into the random access memory when the computer is switched on. Figure 1.8 shows a typical small computer configuration, with CPU, memory, two disc drives (one fixed, one floppy),

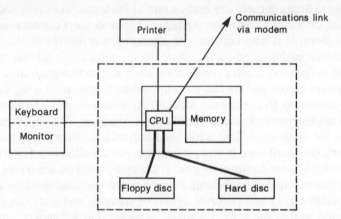

Figure 1.8   Schematic representation of a single-user microcomputer with a CPU, random access memory, floppy and hard discs, printer and visual display terminal (VDT). The VDT consists of a display screen and a keyboard. The modem links the computer to a network via telephone lines. The items enclosed by the dotted line are located within the processor box, which is normally cooled by a fan. The printer, VDT and modem are said to be *peripherals* because they are located outside the processor box.

a printer, a monitor and a keyboard. A larger computer would have a more powerful CPU, a larger memory, more external storage devices, and a more comprehensive operating system capable of handling several users simultaneously. A VAX computer, for example, can handle up to 70 or 80 terminals as well as control other devices such as graphplotters and digitizers for the output and input of map data. Graphplotters and digitizers are described in Chapter 4. Other devices that can be linked or interfaced to a computer are described in Chapter 5, which deals with the topic of remote sensing. Remotely-sensed images can be displayed using an image-display subsystem, which is capable of holding an image in numerical form and converting it to a TV input signal.

1.7.1 Applications software

The computer is not complete without software. Operating system software has already been described, but a computer with operating system software only is not a great deal of use. It requires programs, called *applications software*, which will provide facilities to carry out the operations that the user requires. Examples are word-processing, database management, cartographic data handling, remote sensing, and statistical analysis. Many users prefer to buy "packaged" programs that will carry out these operations, rather than write their own. Examples of packages are Wordstar (word-processing), dBase (database manipulation), ARC/INFO (spatial data handling), SYMAP and SYMVU (map and graph production) and SPSS (statistical analysis). Two of these packages, Wordstar and ARC/INFO, are described below. SYMAP and SPSS are used as examples in later chapters.

1.7.1.1 Wordstar

The Wordstar word-processor is available in a form suitable for most microcomputers running the MS-DOS, CP/M and CCPM operating systems. It is perhaps the most widely-used word-processor among microcomputer users. A word-processor is a program that allows a user to enter text from a keyboard, edit errors, insert, delete and move blocks of text and save the result in a disc file. More advanced word-processors incorporate a spelling checker and a thesaurus. Word-processors such as Wordstar can be used to enter program instructions and data as well as English-language text into files on a fixed or floppy disc. Some larger computers, with operating systems such as RSX-11M, VMS and UNIX, provide an editor utility which is designed for entry and correction of program instructions and data. For example, the VMS operating system provides a utility called EVE (Extended VAX Editor) for these purposes. However, many users of personal computers running MS-DOS use Wordstar, or another word-processor, to perform these functions.

In its basic form, Wordstar is called up by the command "ws". Once the opening menu appears on the screen the user can decide whether to work on text held in an already existing file, create new text (to be stored later) or print the contents of a file previously prepared by Wordstar. Text is entered normally by typing at the keyboard, although it is not necessary to press ⟨return⟩ at the end of a line. Wordstar will left- and right-justify (line up) text automatically, if required. In the justification mode, the ⟨return⟩ key is pressed only to indicate the end of a paragraph.

Wordstar has a large number of built-in commands, providing facilities for the user to delete characters, words and lines, move or copy blocks of text, find and replace words or phrases, change from one print style (e.g. **boldface**) to another (e.g. *italic*), print superscripts and subscripts, and save text to a named file. Unlike more advanced word-processors, Wordstar is not a "WYSIWYG" (what you see is what you get) word-processor. Some word-processors show you on the screen exactly what will appear on the printed page. Wordstar does not; it uses control codes (such as CTRL B, which is generated by pressing the CTRL and B keys simultaneously) to indicate a particular operation (print in boldface type). More up-market versions of Wordstar and other, more expensive, word-processors such as WordPerfect, can simulate the appearance of the printed page on the screen. The Mackintosh range of micros runs possibly the best available WYSIWYG word-processor, which offers a variety of different print styles (fonts) all of which can be viewed on the screen before printing.

Wordstar can be used to send the contents of a text file to a printer. For camera-ready text, which is ready for printing or publication, a printer resembling an electric typewriter can be used. This gives typewriter ("letter quality") output, but is slow. The print-head must be changed if special characters are to be printed. An alternative to this *daisy-wheel* type of printer is a laser printer which produces high-quality output but is more expensive than the daisy-wheel printer. A laser printer for a microcomputer currently costs upwards of £2000. For draft-quality printing, at reasonably high speeds, a dot-matrix printer is most often used. Some dot-matrix printers can operate in a near-letter-quality (NLQ) mode that gives output not discernibly different from that generated by a typewriter. Dot-matrix printers range in price from £150 upwards, though quality and facilities depend on price. A more detailed discussion of printer technology can be found in section 4.2.2.2.

1.7.1.2 *ARC/INFO*

ARC/INFO is a software package used for creating, managing, analysing and displaying geographical data. It generally runs on a large minicomputer such as a VAX, though versions exist that will run on the IBM PC/AT range of micros using the MS-DOS operating system, with a minimum of 30 Mb hard disc and 1.4 Mb floppy. Because of the considerable computational load generated by the manipulation of a large cartographic database the computer should be fitted with a floating-point accelerator or maths co-processor. This co-processor speeds-up arithmetic operations. A digitizer and a plotter (Section 4.2) are also needed. The digitizer reads coordinates from maps, so that the map can be stored in numerical form. The plotter is used to draw maps from their digital representation. A high-resolution graphics monitor is required to provide for the output of maps to the screen.

The ARC segment of ARC/INFO is a graphics data entry, manipulation and display module. It allows the user to input map coordinates of points and lines, using the digitizer, and to store these digitized maps on disc. These maps can be corrected and updated as required, using an editor. Other information (such as boundaries from soil or geological maps) can be added to the topographic base map in the form of an overlay of polygonal boundaries (described in more detail in Chapter 7). Sophisticated data manipulations (such as defining buffer zones of a selected width around linear features such as rivers) can be performed, and map products created and output, either to the monitor or to the plotter. The INFO module is a database management system. The database is made up of information relating to the geographical points, lines and areas shown on maps which have been input using ARC. This information would normally consist of tables of demographic, climatic and other data referring to areas (such as administrative regions), lines (such as roads, rivers and railways) or points (for example, individual buildings or sites of rainfall or river gauging stations). INFO allows the user to issue commands like: select all administrative regions having a total population greater than 500 000 with at least 40% of the population being aged 65 years or more, and having an average per capita income of at least £4000. The administrative regions selected on the basis of this criterion could then be displayed on a map, using ARC.

ARC/INFO and Wordstar are examples of applications programs that are widely-used by geographers in the collection, analysis and display of data and the presentation of results. There are many other applications programs available; some are listed above. Your computer centre should be able to give you a list of all applications programs available locally, or via a network (see below). Three examples of applications programs are used in later chapters of this book – SPSS, the Statistical Package for the Social Sciences, in Chapter 3 and two computer mapping packages, SYMAP and GIMMS, in Chapter 4.

1.8 THE FUTURE

Technological developments in the field of computing are occurring at a rapid rate. Three such developments described here are parallel processing, networks and fifth generation computers.

1.8.1 Parallel processing

Most computers can only execute one instruction at once. If you have a problem that involves the same instruction being executed on each element of a large dataset then the instruction will be carried out serially, that is, on one data element at a time. If it were possible to carry out the same instruction simultaneously on a set of data elements then the entire operation could be performed very much more rapidly. A parallel processor can do just that. Each of a number of individual slave CPUs performs an operation on a piece of data and returns the result to a master CPU. It does not take much imagination to realize the implications of parallel processing for many of the geographical operations described in this book. In Chapter 5 the subject of digital processing of satellite images is considered. Each satellite image is made up of a set of horizontal rows of data, with each data element representing a point on the ground. If the average value of the numbers representing the image was required then a conventional serial computer would start at the top left of the image, and add up the numbers one by one along each row. A parallel processor could use a number of slave processors and instruct each one to add up the numbers along one row. These additions could be going on simultaneously so the entire operation would be completed far more quickly than would be the case if a serial processor was used. At the time of writing (1990) it is possible to buy parallel processing machines based on the Inmos Transputer. However, a great deal needs to be done to make them generally usable, but when that day arrives even a desktop computer will be able to outperform many of the large serial machines in use today.

1.8.2 Networks

Networks consist of a number of computers joined together by tele-communications links, which may be physical telephone cables or radio links, perhaps even using communications satellite channels. A computer linked into the network can access programs and data held by any other computer in that network. If the telephone network is being used then the signal sent from one computer to another passes through a device called a *modem*, which makes the signal suitable for transmission through the telephone system, or converts a signal received through the telephone system into computer-readable form. Computers in the same building or on the same campus can use a dedicated high-speed network system, the best-known of which is Ethernet. Networking will allow access to data and programs held on other machines; for example, computer A may hold the census data for an area and the programs needed to access and analyse that data. Computer B may hold programs for automated mapping. A user who has access to a small computer at site C may want to analyse the census data at A and produce maps of the results using the facilities at site B.

If A, B and C are all on the same network then the user at C can do exactly what he or she wants.

Networks give access to computing power and to data archives stored at geographically-remote locations. Networks can, of course, also be used to access a nearby computer. Many campuses have *Local Area Networks* or LANs which link together all the publicly available computers on the campus. A student can access any of these computers from a single terminal. At the other extreme, multinational corporations link together their computer operations to allow any of their national or regional offices to access the computer facilities available at any other national or regional office. Each office has access to all the information in the network. In addition, idle computer time can be used, since time differences across the world mean that office staff are starting work in Los Angeles at about the same time that their equivalents in London are going home.

Universities and research institutes in the UK are connected via a network called the Joint Academic Network (JANET). This network allows users to access remote computers as described above; it also permits individuals at different sites to communicate using *electronic mail*. A user at site A can type a message on his or her terminal and send it, via JANET, to a user at site B. JANET is linked into the European Academic Research Network (EARN) and into other networks, such as Bitnet (which operates in the USA). The availability of these networks means that research workers in the UK, other European countries and the USA can communicate easily and share their programs and databases more readily.

1.8.3 Fifth generation computers and expert systems

Fifth generation computers are a matched or balanced amalgam of hardware and software with access to large volumes of data (databases). The information technology revolution is made possible by the application of this composite of hardware, software and data to real-world problems. Besides numerical records of, for instance, well logs or rainfall values a database could also contain information provided by experts (the knowledge base). New programming languages such as PROLOG allow the computer to simulate the process of inference, so that problems can be approached using both the analytical capabilities of the computer and the information stored in its knowledge base. This process is termed *Artificial Intelligence* (AI). One type of Artificial Intelligence system is called an Expert System. Expert Systems provide access to the knowledge and experience of specialists in a particular field. This knowledge is used in particular cases where the user might be inexperienced or unsure. A doctor with access to a medical diagnosis Expert System could input the details of a patient's symptoms. The medical Expert System would then ask questions which would allow it to infer from the information provided by the doctor and from the knowledge stored in its knowledge base whether or not the diagnosis was correct. The Expert System thus consists of

- A general knowledge base, possibly expressed in the form of rules such as "if two lines are parallel then they will never cross". These rules, which are called production rules, do not relate to particular examples but to general problems. They need not be certain, as in the example cited, but may result in a probable conclusion, for example: "if the barometric pressure is less than 990 millibars and if it is cloudy then the probability of rain is 0.75".
- Specific information related to the problem to hand (for example, observations of barometric pressure, cloud cover, and other meteorological measurements made on a particular day).
- Procedures which allow deductions to be made from the application of the general knowledge base to the specific, problem-related data.

Expert Systems could be used within geography in education, training and research applications. The use of statistical methods, for example, requires a wide-ranging knowledge covering the nature of the techniques and the assumptions on which they are based. Many geographers are familiar with statistics but are by no means specialist statisticians. If the knowledge of professional statisticians could be made available to them through a friendly Expert System then the use of statistical methods could proceed not on a trial and error basis but through a dialogue between the user and the Expert System. Another example might be map design and production. If geography students were able to call upon the resources of a cartographic Expert System (perhaps via a campus network) then they might learn the principles of map design more quickly and also be able to produce high-quality computer-drawn maps more easily.

The ideas described above are, with the exception of networks, still at the development stage. There is no doubt, though, that as research proceeds the face of computing in the 1990s will be changed in a quite dramatic way.

Additional reading covering material presented in this and later chapters is provided by McGuire (1989).

1.9   REVIEW QUESTIONS

1. Explain the meaning of the following terms:

hexadecimal	base 10	binary
byte	bit	CPU
floppy disc	RAM	operating system
ROM	file	parallel processing
network	Expert System	hardware
applications software	program	ASCII

2. In what ways do you think that geographical users of computers will benefit from developments in (i) networks and (ii) parallel processing?

3. Explain how a computer processes a program written in either the BASIC or FORTRAN languages. Why are programs written in high-level languages such as FORTRAN and BASIC to be preferred to assembler language programs?

4. Why is an operating system necessary? Describe how you would use the operating system commands of a computer with which you are familiar to:

- list the disc files you own

- create a new file or amend an existing file using an editor

- run a program from an applications software package such as SYMAP or SPSS.

CHAPTER 2

Computers and Geographical Data

2.1 INTRODUCTION

The important properties of geographical data are considered in this chapter. Geographical data consist of measurements of the location of objects on the surface of the Earth and of the properties (attributes) of such objects. The surface of the Earth is considered to be two-dimensional with the position of an object being given by a horizontal coordinate (x) and a vertical coordinate (y). Interest in geographical or spatial data centres not only on the location of the object described, and its characteristics, but also on the relative location as measured by such properties as connectivity and adjacency. These latter are called *topological* properties.

The topics considered in this chapter are

- the nature of geographical data, both locational and attribute, and
- the structure of geographical databases.

The term *dataset* is used to mean an assemblage of data describing specific attributes of an area. Individual datasets may consist of measurements of the characteristics of climate, demography, or transport networks. The term *database* is used to mean a collection of related datasets. The *structure* of a dataset is the way it is considered to be organized. An example of a structure is a row and column organization in which the rows contain data for a particular observational unit such as a town, a county or a point while the columns contain the measures over all observational units for a single variable such as rainfall, population density or soil type. Appendix A contains an example of such a data structure. Note that the physical memory of a computer is made up of a sequence of byte locations, labelled upwards from 0. The row/column organization is one which is assumed by a programmer or user and is therefore said to be a logical (rather than physical) structure. A more complicated data structure would have multiple datasets with pointers connecting the elements of one dataset with those of another. It is the nature of the data, and its structure, which effectively determines the use or range of uses to which those data can be put.

30

Geographical data are of two kinds. The first kind consists of values which describe the *attributes* of points, lines or areas. These values can be labels, ranks or magnitudes; for example, the mean annual rainfall for Cheshire is a value associated with an area, while the height above sea-level of the summit of Ben Nevis is a value associated with a specific point. The second type of geographical data relate to the specification of *locations*, not to the properties of entities at those locations. Location coordinates, such as latitude and longitude, can be used to define the position of a single point, or they can be used to define the position of points located along a line, such as a contour line or a boundary line. In turn, the lines can be used to delimit areas. Thus, geographical data consist of values describing the properties of objects (points, lines or areas) or defining the location of those objects. This distinction is not always made, but it is useful for ordering and therefore accessing spatial datasets.

2.2 SCALES OF MEASUREMENT

There are several ways of classifying the numerical data that are used in geography. The approach followed here is based on the concept of the scale of measurement, for this approach emphasizes the information content of different levels or types of data that can be utilized in geographical study. The differences between the data types are not just of academic interest for they determine the kind of question that can be addressed to the data. They also define the statistical and numerical methods that are appropriate in the answering of such questions. The scale on which data are recorded is an inherent property of data. The use of the term *scale* here should not be confused with the scale of a map. Scale in the context of this section is a property of numerical data.

There are four recognized scales of measurement on which data may be recorded: nominal, ordinal, interval and ratio. Nominal is the lowest level of measurement and ratio the highest in terms of information content. The interval and ratio scales of measurement will be considered together, so three scales of measurement will be identified. An example illustrates the difference between the scales. Table 2.1 shows the mean monthly temperatures during the 12 months of the year (January to December) for Thessaloniki, Greece. Column 1 of Table 2.1 shows the temperatures expressed on a nominal scale. The symbol 1 is used merely as a label; it means that the temperature is 60°F or more, whereas the symbol 0 means that the temperature for that month is less than 60°F. Neither the actual temperature values nor the differences in temperature from month to month can be obtained from the data in column 1. If these data are used the only questions that can be answered are those such as "How many months of the year have an average temperature of 60°F or more?", or "Is the number of months with an average temperature less than 60 greater than the number with an average temperature of 60°F or more?". If such questions are the only ones likely to be asked of the data then the use of the nominal

Table 2.1 Mean monthly temperature for Thessaloniki, Greece, expressed on the nominal, ordinal and interval scales. The interval scale measurements are in °F

Month	Nominal scale	Ordinal scale	Interval scale
January	0	12	45.0
February	0	10	48.0
March	0	8	55.0
April	1	7	63.0
May	1	5	70.0
June	1	3	77.0
July	1	1	80.0
August	1	2	78.0
September	1	4	74.0
October	1	6	65.0
November	0	8	55.0
December	0	10	48.0

scale is a suitable and economical way of recording such data. Notice that more than two labels can be used; the digits 0 and 1 are used in column 1 of Table 2.1, but for other purposes it would be possible to use more than two labels. Labels are non-negative whole numbers such as 0, 1, 2, 3 or 4.

If label values are expressed in base two form, as explained in Chapter 1, it will be seen that the amount of memory storage that is taken up by the data is dependent only on the number of levels used in the nominal scale. If only two levels are used then one binary digit (bit) per value will be needed, since a binary digit can take one of two states, 0 or 1. The values for eight observations can therefore be held in one byte of computer storage. If the number of levels is increased to three or four then two bits will be needed for each observation, giving the four possible codes 00, 01, 11 and 10. The number of bits per observation that are needed can be worked out by expressing the number of levels as a binary number (Table 2.2). The formula used to determine the numbers in the first column of Table 2.2 is: two to the power of the number of bits required, less one, that is, $2^n - 1$. Since 2^1 is 2, 2^2 is 4 and 2^{10} is 1024, one bit is needed if the number of levels is one, or two bits are needed if the number of levels is three or four, and 10 bits are needed if the number of levels is between 513 and 1024. The storage requirements for nominal scale data are small relative to the requirements of ordinal and interval/ratio scale data.

The second column of Table 2.1 shows the same temperature data expressed on an ordinal scale. The ordinal scale brings in the concept of order or ranking, but does not maintain the actual data value such as 60°F. The ordinal scale allows new questions to be asked, for example, "Which is the hottest month of the year in Thessaloniki?". This kind of question cannot be asked if the data are measured on a nominal scale. However, ordinal scale data will provide answers to all the questions that could be asked of nominal scale data. In the

Table 2.2   Number of bits required to represent base 10 numbers 0–1023

Base 10 number	Base 2 number	Number of bits
1	1	1
3	11	2
7	111	3
15	1111	4
31	11111	5
63	111111	6
127	1111111	7
255	11111111	8
511	111111111	9
1023	1111111111	10

The table shows the largest positive number that can be held in a memory element of a given size, measured in bits. Thus, for example, a 10-bit memory element can store positive (base 10) numbers in the range 0–1023 whereas eight-bit memory elements cannot hold positive numbers larger than 511. In most computer systems the available range is split into two with half the range reserved for negative numbers; an eight-bit memory element (a byte) is generally used to hold positive numbers from 0 to 127 and negative numbers between −1 and −128.

example notice that March and November have the same average temperature so they are both given rank 8. Hence there is no month ranked 9.

The size of the individual data storage unit for ordinal scale data depends on the number of individual pieces of data. In the example 12 individual measurements made up the dataset, so the highest possible rank is 12. If the states of the USA were ranked then the maximum rank value would be 50. If the number of individuals is less than 128 then a single byte can be used to hold an ordinal scale value, because the maximum positive number that can be stored in eight bits is 127. If the maximum rank is greater than 127 but less than 32 767 then a two-byte storage unit must be used for each item. A four-byte storage unit can hold positive integer numbers, up to 2 147 483 647. These limits depend on the kind of computer used and also on the programming language. The values given here refer to a byte-oriented computer and a language such as BASIC or FORTRAN, as described in Chapter 1.

The data in column 3 of Table 2.1 are the actual mean monthly temperatures at Thessaloniki. They are expressed on an interval scale, which means that arithmetic differences are comparable and do not depend upon position along the scale. Look at column 2 of Table 2.1. The concept of "May temperature minus June temperature" is meaningless – though the misguided might be tempted to answer "two". Using the interval scale representation, one could correctly answer 7°F to the same question. The difference of 7°F does not depend on the fact that the two months concerned are May and June. Incidentally, the ratio scale, which was mentioned earlier without explanation, is a refinement of the interval scale in that the data have a physically-defined, as opposed to an arbitrary, zero point. A ratio scale of temperature might have its zero point defined by the freezing point of water (as the Celsius scale does) or by absolute zero ($-273°C$) as the Kelvin scale does. Both the ratio and the interval scales contain the information that is expressed in the nominal and ordinal scales. It is evident from inspection of column 3 of Table 2.1 which months have temperatures above or below 60°F. The months can also be ranked in order of decreasing temperature. In addition, answers to questions such as "What is the temperature difference between September and May?" or "Is that difference less than the difference between January and April?" can also be found.

The extra cost in computer storage required when data are expressed on an interval/ratio scale instead of a nominal or ordinal scale can be easily calculated. Interval/ratio scale data are generally in the form of decimal numbers such as 45.0, 33.969 or -19.22. This kind of number is stored in "floating-point format" which typically uses a four-byte storage unit per number. If more precision is required then an eight-byte unit can be used. The numbers listed in column 3 of Table 2.1 can be stored in four-byte units, which means that the number of bytes required to hold all 12 numbers is 48. At the ordinal scale only 12 bytes would be required, whereas at the nominal scale only 12 bits or one and a half bytes would be needed.

2.3 LOCATIONAL DATA

So far, the data that have been used as examples of the three different scales have not been inherently geographical. Although these data refer to a place (Thessaloniki) no measured element of location is involved. In this section the way in which a spatial or locational component can be associated with sets of non-spatial data is considered, and ways in which such locational data can be structured for computer manipulation are examined.

Spatial data can take one of three forms which are, in fact, levels of a hierarchy. The first is point data. The second, line data, consists of a set of points while the third, areal data, is defined in terms of an intersecting set of lines which form a closed boundary. Sometimes the area within this closed boundary is called a polygon. Figure 2.1(a) shows the three spatial data types.

Point data can be of two kinds. One is actual measurements at particular geographical locations. The values of rainfall measured at a set of rain gauges would constitute a set of point data linked to individual locations. The second type of point data does not really refer to points at all but to areas. These derived data are, in fact, averages of spatially-distributed values that are, for convenience, labelled as point data. For example, the populations of a set of counties might be represented on a map by a set of points, each point located at the areal centre of its county. A contour map might then be drawn on the basis of these point values to show the variation in population. However, this

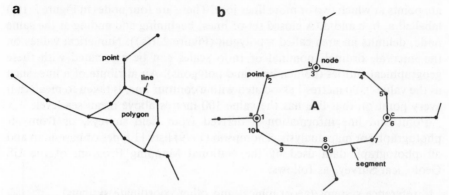

Figure 2.1   (a) Point, line and polygon spatial data types. A line is made up of a set of points while a polygon is delimited by a set of intersecting lines. Polygons may also be called zones or counties. (b) The boundary of the polygon representing county A is defined by four lines. Each line joins two nodes (open circles). A node is a point lying at the junction of two or more lines. Segments join two points. Lines are composed of sets of contiguous segments.

implies that all the people in a particular county live at the centre of gravity of the area of that county, which is rarely if ever the case. The contouring operation might not be done explicitly but via a statistical mapping technique such as Trend Surface Analysis (Section 3.4.3); the users of such techniques should be aware of the true nature of their data so that unwarranted or irrational inferences are not drawn.

Point data have three kinds of attribute:

● magnitude or value, expressed on one of the measurement scales discussed earlier;
● location or position, in terms of a coordinate system such as latitude and longitude, or eastings and northings on a grid; and
● time, either absolute time (such as: AD 1999) or relative time (such as: 10 years ago).

The data available for a geographical point can thus be expressed as a value at a location at or during a particular time. It is these characteristics which are used by geographers to impose a structure on the data with which they are working; data that are not structured or arranged in an orderly fashion are rarely of value, for structure is needed if information is to be extracted from data.

The locational attributes (x,y coordinates) of points can be used to define other spatial entities. Figure 2.1(b) shows 10 points located along the boundary of county A. Pairs of spatially-adjacent points are joined by segments, and the segments link together to form lines. A line joins two nodes, whereas a segment joins two points which are not nodes. Nodes are circled in Figure 2.1(b). The points are usually located at places where the boundary alters direction. Nodes are points at which two or more lines join. There are four nodes in Figure 2.1(b) labelled a, b, c and d. A closed set of lines, beginning and ending at the same node, delimits an area called a polygon (Figure 2.1(a)). Numerical values on the interval, ordinal, nominal or ratio scales can be associated with these geographical entities (points, lines and polygons). An attribute of a line, such as the value "100 metres" associated with a contour line, is taken to mean that every point on that line has the value 100 metres above mean sea-level.

Point and line information is derived from field surveys, or from air photographs or map analysis. Thompson (1979) lists 11 types of base-map and air photograph data used by the National Mapping Program of the US Geological Survey, as follows:

1. reference systems (geographical and other coordinate systems)
2. hypsography (contours, elevations and slopes)
3. hydrography (lakes and rivers)
4. surface cover
5. non-vegetative features
6. boundaries (political and administrative)
7. transportation systems
8. other significant manmade structures
9. identification and portrayal of geodetic control
10. geographical names, and
11. orthophotographic imagery.

Some map data are already available in computer-readable form. Other data must be extracted from the paper map itself; this is an error-prone and time-consuming operation if carried out by hand. Mapping agencies and research institutes use machines which assist in the process of converting map and photographic data into digital form. These machines are called digitizers (Figure 1.3), and the cheapest version is simply a table on which the map or photograph is placed. A cursor is used either to indicate the location of a point or to trace along a line. The position of the cursor is recorded whenever a button is pressed, though some models will record the cursor position at regular time intervals. This feature is useful for line-following work. The user can also add textual data associated with points (for example, the name of a town or of some other feature) using a keyboard. Data from the digitizer are fed directly into the computer and are subsequently stored in a disc file.

More automated methods of digitizing are available though naturally the equipment is more expensive. These methods include the interactive laser system which involves the projection of the map or photograph onto a large screen. The lines indicated by the operator are digitized automatically using advanced line-following techniques. Another automatic method is called raster encoding. The map or image is scanned by a beam of light and the reflectance (level of grey) of the map or photograph at each of a large number of points is recorded.

Each point represents a very small rectangular map area with which a numerical value is associated. The value indicates whether the cell is black (0), mid-grey (127) or white (255) or somewhere in between. Line-following methods are again used to extract the coordinates of points lying along specified lines. Neither method can detect whether a line is a contour line and if so, the height it represents, or a road or a stream. This information, which is sometimes called intelligence, has to be added by a human interpreter after digitizing is complete. A raster-encoded (digitized) version of the letter I is shown in Figure 2.2. Digitizing methods are described more fully in Chapter 4.

A considerable amount of raster-encoded data is directly available to geographers in the form of digital imagery from Earth observation satellites such as Landsat, SPOT, TIROS/NOAA and Meteosat. These satellites carry instruments which scan the Earth's surface and record the reflected sunlight or emitted heat from each of a large number of small ground areas. In North America the data are available from the EOSAT Corporation in the case of Landsat and from the National Oceanographic and Atmospheric Agency (NOAA) for meteorological satellite data. In the United Kingdom, satellite imagery can be purchased from the National Remote Sensing Centre at Farnborough, Hampshire. Remotely-sensed data are described in detail in Chapter 5.

2.4 DATA STRUCTURES FOR GEOGRAPHICAL DATA

The logical (as opposed to physical) way in which a dataset is considered to be organized is its structure. Different data structures are appropriate for different applications and for different types of data. The way in which a dataset consisting of street names, property types and locations in a town is structured will depend in the first instance on its format (raster or vector, as described below) and on the use to which the data are to be put. If the data are to be

```
000000000000
000111111000
000001100000
000001100000
000001100000
000001100000
000001100000
000001100000
000001100000
000001100000
000111111000
000000000000
```

Figure 2.2 The letter I in raster format. The character 0 represents a white background. 1 represents black foreground.

used for determining the optimum route between two points in the town then the data structure should allow street junctions to be accessed rapidly, with pointers embedded in the data structure to direct the search from one junction to the next. A data structure based on the alphabetical order of the street names could not be efficient for this problem, as the database would have to be searched each time a street intersection was required. Other problems require other structures, and one of the difficulties in storing geographical and associated data is to decide upon the most flexible and efficient way to structure the dataset. In this section some introductory aspects of the topic of data structures for raster- and vector-coded datasets are considered.

2.4.1 Raster data and the quadtree

Raster datasets can be envisaged as sets of two-dimensional matrices, with the cells of each matrix representing rectangular areas on the ground and containing a code which may be an ordinal-scale number or simply a label. A remotely-sensed raster dataset (considered in Chapter 5) consists of several registered matrices or two-dimensional rectangular data structures arranged in row and column order with the values in each of the cells of the matrix measuring the sunlight reflected, or the heat emitted, by each of a large number of small areas of the Earth's surface. Matrix 1 might represent the amount of visible green light that is reflected by each cell, while matrix 2 might represent the reflectivity of the same set of ground areas in the red region of the spectrum. The reflectivity values are represented on an ordinal scale ranging from 0 (which may or may not mean zero reflectivity; it merely represents the lower limit of the detector's capability) to an upper value of 63, 127, 255 or 1023 depending on the

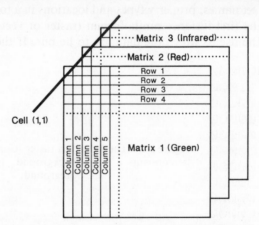

Figure 2.3 Row/column/waveband registered raster dataset. The four components of this dataset represent the reflectivity of ground elements in the green, red and two near-infrared bands of the spectrum (Chapter 5).

sensitivity of the detector. Because of the great variability in reflectance over the Earth's surface it is usual to store each matrix in its raw form. The data structure in this case is a very simple row/column/waveband configuration, as shown in Figure 2.3.

If a raster dataset is processed by classification or pattern recognition methods as described in Chapter 5 the result is a single matrix in which the cells are labelled according to some meaningful scheme; for example, the label '1' might be given to cells which are identified as water, while cells covering ground areas thought to have a bare sand surface might be labelled '2', and so on. Figure 2.4(a) shows an extract from a four-channel registered raster dataset. In Figure 2.4(b) areas of low reflectance in all four channels, thought to be water bodies, are given the label '1' and areas of high reflectance typical of sand are given the label '2'. Despite the fact that the original four-channel data contained significant cell-to-cell variability, the product (the classified dataset) consists of only two regions, one labelled '1' and the other labelled '2'. The information in the labelled dataset can be stored in a more efficient way than is the case for the raw data, for the labelled dataset has only two values and the regions occupied by the two labels are distinct and homogeneous (Figure 2.4(b)).

A hierarchical data structure called a *quadtree* can be used to store raster data which consist of labelled regions whose spatial structure is homogeneous. For example, the pattern shown in Figure 2.4(b) represents an area covered by an 8×8 array of cells. The value in each cell is either 1 or 2, depending on whether the cell is a member of the class "water" or a member of the class "sand". The 8×8 matrix is firstly divided into quadrants, labelled 0, 1, 2, and 3 in the order northwest, northeast, southwest, southeast (Figure 2.5(d)). If any quadrant consists of all 1s or all 2s it is left alone, otherwise the quadrant is split into four sub-quadrants, labelled 0–3 as before. Again, homogeneous quadrants are left alone whereas non-homogeneous quadrants are subdivided into four. This process continues until all remaining quadrants are homogeneous or until the subdividing procedure reaches the individual cell level. Figure 2.4(c) shows the subdivision procedure applied to the dataset of Figure 2.4(a); another way of representing the data structure is in the form of a tree (Figure 2.4(d)). Only the descriptions of the end-points of the branches of the tree need be stored, rather than the entire 8×8 matrix of Figure 2.4(b), and certain operations can be carried out on the data stored as a quadtree in a more economical and efficient manner than would be the case if the data were represented in the row-and-column fashion.

Figures 2.5(a) to 2.5(d) illustrate the essentials of the quadtree structure as well as the simplicity of the addressing mechanism. The cells of Figure 2.5(a) show the row number (above) and the column number (below) for each cell of the matrix. The rows and columns are numbered 0–7 rather than 1–8. Figure 2.5(b) is the same as Figure 2.5(a) except that the row and column numbers are represented as base two numbers. Each row and column coordinate

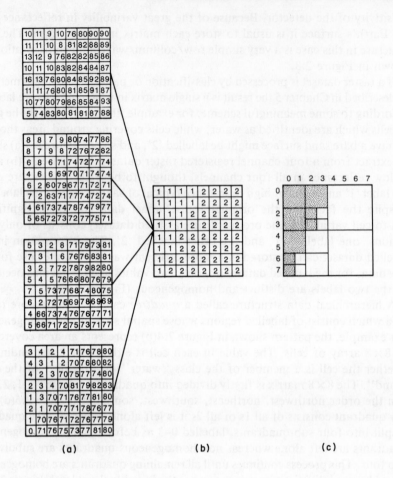

(a)

(b)

(c)

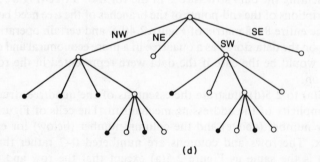

(d)

has three binary digits because the number of rows and columns is eight, which is two to the power of three (see Chapter 1 for a discussion of base two numbers). Quadtree addresses are obtained by reading down each of the three columns of the row and column coordinates and expressing each pair of digits as a base two number then translating the result to base 10. Since that sounds difficult, a simple example is given. In base two form, the value in row 4, column 7 in Figure 2.5(b) (remembering that we are counting the rows and columns from 0) is:

100   (row number: 100 in base two is 4 in base 10)
111   (column number; 7 in base 10 form).

The three columns are read separately, giving the three two-digit values (11), (01), (01). In base 10 notation these three pairs of base two digits are written as 311. The meaning of each digit (0, 1, 2 or 3) is shown in Figure 2.5(d).

This apparently pointless mixing of the digits of the (base two) representation of the row and column coordinates of the raster data structure has produced a method of geographical addressing. If the symbol "∗" is used to mean "all quadrants (or sub-quadrants, or sub-sub-quadrants) then the pattern of water and sand in Figure 2.4(b) and (c) can be coded in an economical way. The quadtree address "∗∗∗" means "all quadrants, all sub-quadrants, and all sub-sub-quadrants" in a three-level quadtree, that is, the entire 8 × 8 matrix. The quadtree address "32∗" means "southeast quadrant and entire southwest sub-quadrant", i.e. the group of cells with row/column coordinates (6,4), (6,5), (7,4) and (7,5). Using this notation the cells occupied by water in Figure 2.4(b) can be written as 00∗ 01∗ 02∗ 030 032 20∗ 220 222 while the cells occupied by sand are described by the string 031 033 1∗∗ 21∗ 221 223 23∗ 3∗∗, using the addressing mechanism described above. The savings in computer storage are considerable, and computer time can also be saved if operations are carried out on the quadtree rather than on the entire raster structure stored in matrix

---

Figure 2.4 *(opposite)* (a) Four-waveband registered raster dataset. The number in each cell is a measure of the reflectance of the surface material at the corresponding point on the ground. (b) Labelled cells show areas recognized as water (1) or sand (2). (c) Areas labelled 1 in Figure 2.4(b) are stippled; blank areas are labelled 2 in Figure 2.4(b). (d) Tree representation of Figure 2.4(c). The root node (top) represents the full matrix shown in Figure 2.4(c). Since the full matrix is not homogeneous it is broken down into four quadrants (NW, NE, SW, SE). The NW quadrant has three homogeneous sub-quadrants (NW, NE, SW) but the SE quadrant is broken down into four sub-quadrants, two of which are black (representing the label 1 in Figure 2.4(b) and two are white (label 2). The NE, SW and SE quadrants can be interpreted in a similar fashion.

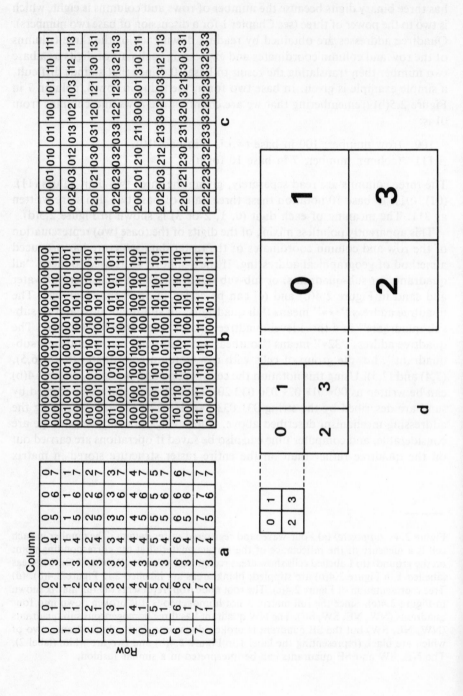

forms shown in Figure 2.4(b). The savings in storage and computing time are strongly dependent on the degree of homogeneity of the raster structure. A pattern such as that shown in the example can be compacted quite considerably, but the same could not be said of any of the four original bands of data shown in Figure 2.4(a). There is a balance between the time saved by operations on the quadtree, the time taken to convert the raster data to quadtree form, and the degree of compaction achieved by the quadtree data structure. For homogeneous patterns the savings are worthwhile but for a heterogeneous pattern the quadtree structure will prove to be less efficient.

Further details of quadtree methods are provided by Burrough (1986) and Hogg and Stuart (1987). The possibility of designing hardware to store and search quadtrees in an efficient fashion is discussed by Oldfield *et al.* (1987).

2.4.2 Vector data

2.4.2.1 Polygon encoding

Vector data consist of strings of $\{x,y\}$ coordinates giving the positions of points which are related in some way. Figure 2.6 shows a simple map which shows a rectangular street pattern. The street junctions are called nodes and are given labels. A simple way to represent the locations of the streets and the four blocks (labelled A to D in Figure 2.6) would be to give the $\{x,y\}$ coordinates of the nodes at street intersections along a given street, or at the boundaries of the blocks. The order in which the nodes are stored does not matter, as long as

---

Figure 2.5 *(opposite)* (a) Cell addresses in an 8×8 raster structure. Row number is uppermost in cell. Note that rows and columns are numbered starting from 0. (b) Same as Figure 2.5(a) except that row/column identifiers are written in base two notation. (c) Quadtree addresses corresponding to Figure 2.5(b). The individual cell address is formed by reading the three vertically aligned pairs of base two digits in each cell of Figure 2.5(b). For example, the cell on row 2, column 1 has three vertical pairs '00', '10' and '01' which translate to 021 since base two '00' is base 10 '0', base two '10' is base 10 '2' and base two '01' is base 10 '1'. See Table 1.1 for details of converting from base two to base ten notation. (d) Quadtree spatial addressing (see also Figure 2.4). The full matrix (right) is decomposed into quadrants (labelled 0, 1, 2, 3). These quadrants are split into sub-quadrants and further into sub-sub-quadrants until the individual pixel level is reached. Figure 2.5(c) can be subdivided three times before the individual pixel level is reached. At the first level the quadrants are 4×4 pixels, at level two they are 2×2 pixels and at level three they are 1×1 pixels. The three levels are represented by the three digits in each cell of Figure 2.5(c). The first digit tells us which level one quadrant in which the cell lies (with 0, 1, 2, 3 representing NW, NE, SW, SE), the second digit gives us the second level quadrant and the third digit gives the level 3 quadrant. Thus, the label of the lower right cell in Figure 2.5(c) is 333. The first 3 says that the cell is in the SE quadrant at level 1, the second 3 says it is in the SE quadrant at level 2, and the third 3 says it is in the SE quadrant at level 3.

the scheme is consistent, that is, if the nodes lying on the boundary of block A are presented in clockwise order then all other block boundary nodes must be stored in clockwise order. Thus, block A would be represented by the string of coordinates $\{x_1,y_1\}$, $\{x_2,y_2\}$, $\{x_{10},y_{10}\}$, $\{x_8,y_8\}$, $\{x_9,y_9\}$, and, depending on the way the program was written $\{x_1,y_1\}$. If these symbolic coordinates are read off the map we get the string of numbers 5,60, 45,60, 45,30, 35,30, 5,30 (,5,60). The $\{x,y\}$ coordinate string for block B would be: 45,60, 70,60, 70,30, 45,30 (,45,60). This scheme is called polygon encoding, for each polygon (block) is digitized separately.

Polygon encoded data are useful if the application to which they are to be put is not dependent on knowledge of neighbouring polygons. Such knowledge is called topological information. For example, if all the blocks have associated value labels such as the '1' and '2' representing water and sand in the example in Section 2.4.1, then it would be comparatively easy, using modern digital display equipment such as that described in Chapters 4 and 5, to respond to queries such as "colour in yellow all the polygons labelled '2'". However, it would be less easy to respond to queries such as "colour in yellow all polygons labelled '2' provided that at least one polygon labelled '1' has a common boundary with the polygon labelled '2'" if polygon encoding were used. Furthermore, digitization is an inherently error-prone procedure, and the separate digitization of the nodes surrounding each polygon might result in the generation of slightly different coordinates for a given node each time it was digitized. The resulting "sliver polygons" could cause trouble if any form of processing was carried out on the digitized data (Figure 7.4).

2.4.2.2 DIME structure

A second way of organizing or structuring the data shown in Figure 2.6 is described by Cooke (1987). This is the DIME (Dual Independent Map Encoding) scheme, developed by the US Census Bureau in 1967. The DIME method of encoding data requires, first of all, that the $\{x,y\}$ coordinates of the nodes are stored separately in a table such that entry n in the table contains the $\{x,y\}$, coordinates of node n. Each line segment, defined as a line joining two nodes, is coded separately with (as a minimum) the node identification numbers and the labels of the blocks to the right and to the left being stored. The determination of what is "left" and what is "right" is decided by looking in the direction of the second node from the first. Thus, the entries

REGENT STREET   8   9   A   C

and

REGENT STREET   9   8   C   A

are equivalent, since block A lies to the left when travelling from node 8 to node 9 whereas block B lies to the left when travelling from node 9 to node 8.

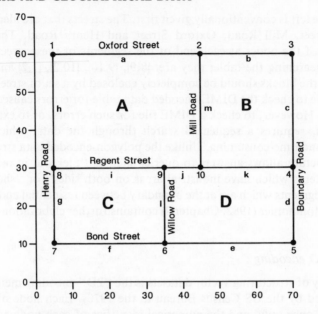

Figure 2.6 Rectangular street pattern showing blocks (A–D, nodes (1–10) and street segments (a–m). See text for elaboration.

Other information, such as street numbers along the segment, postcode, or census enumeration district identifier is stored alongside the node and block information. The full DIME encoding for the map in Figure 2.6 is:

REGENT STREET   8   9   A   C
REGENT STREET   9   10   A   D
REGENT STREET   10   4   B   D
MILL ROAD   10   2   A   B
OXFORD STREET   1   2   99   A
OXFORD STREET   2   3   99   B
BOUNDARY ROAD   3   4   99   B
BOUNDARY ROAD   4   5   99   D
BOND STREET   5   6   99   D
BOND STREET   6   7   99   C
HENRY ROAD   7   8   99   C
HENRY ROAD   8   1   99   A
WILLOW ROAD   6   9   C   D

The code 99 is used here to indicate an unlabelled block. Given the DIME-encoded data it is possible to extract other information. For example, to find the names of the streets surrounding block A simply find all entries 'A' in the columns "block to the left" and "block to the right", remembering that the block to the left is conventionally given first. The streets that are identified are: Regent Street, Mill Road, Oxford Street and Henry Road. The numeric identifiers of the nodes at each end of the segments of these streets are also found by searching the table; they are: {8,9}, {9,10}, {10,2}, {1,2} and {8,1}.

Because the blocks should be completely enclosed by a set of street segments it is possible to check the DIME-encoded data table for errors caused by faulty data entry. However, to check a DIME file for such errors, or to extract block boundaries, requires a sequential search through the data, which may be expensive and time-consuming. Unlike the polygon-encoded data structure, the DIME structure allows answers to queries such as "pick out in red all those street segments which have industrial areas on both sides" or "show in blue the street segments which are at the boundary between residential and industrial blocks". Monmonier (1982, Chapter 7) contains further elaboration of DIME.

2.4.2.3 2D encoding

A third way of structuring vector datasets is the "2D" encoding method which was adopted by the US Census Bureau in the 1970s. Each node on the map is digitized once only, and the numerical identifier of each node and its {x,y} coordinates are stored in a node file, sometimes called a 0-cell file, the 0 referring to the fact that a point has a dimension of zero. A second file, the line or 1-cell file, stores the definitions of the line segments and associated topological information, such as the identifiers of the blocks to the left and right, as in the case of the DIME file. The streets themselves, as well as the street segments, are given identifiers. The block file (2-cell file) is a record of the block identifier on the map, the actual block number and a pointer to the 1-cell (line segments) file. The final file is a street name file, relating street identifiers and street names.

The 0, 1 and 2-cell files for the map shown in Figure 2.6 are listed in Table 2.3.

2.4.2.4 USGS digital line graph

The US Geological Survey produces digital cartographic data which are similar in structure to the 2D encoding scheme described in Section 2.4.2.3. The data are divided into four files, containing respectively

- The Digital Line Graph (DLG) file, containing line information from the map (transport network, hydrography and boundaries) at a scale of 1:100 000 or 1:2 000 000;
- digital elevation model file, containing data relating to elevation above sea-level derived at a 1:250 000 scale;
- land use and land cover file at 1:250 000 and 1:100 000 scales; and
- geographical names.

Table 2.3 Cell files for the map shown in Figure 2.6

0-cell file (node file)

Node number	Coordinates x	y	Pointers to segments
1	5	60	a,h
2	45	60	a,b,m
3	70	60	b,c
4	70	30	c,k,d
5	70	10	d,e
6	35	10	e,l,f
7	5	10	f,g
8	5	30	g,i,h
9	35	30	i,l,j
10	45	30	j,m,k

1-cell file (line segment file)

Segment number	Street id	Node 1	Node 2	Block left	Block right
a	1	1	2	5	1
b	1	2	3	5	2
c	5	3	4	5	2
d	5	4	5	5	4
e	4	5	6	5	4
f	4	6	7	5	3
g	6	7	8	5	3
i	3	8	9	1	3
j	3	9	10	1	4
k	3	10	4	2	4
l	7	6	9	3	4
m	2	10	2	1	2

2-cell file (block file)

Block number	Block id	Pointer to segment
1	A	a
2	B	b
3	C	f
4	D	d
5	99	a

Table 2.3   *(continued)*

Street names file		
Street number	Street name	Pointer to segment
1	OXFORD STREET	a
2	MILL ROAD	m
3	REGENT STREET	i
4	BOND STREET	e
5	BOUNDARY ROAD	c
6	HENRY ROAD	g
7	WILLOW ROAD	l

The DLG file holds data for three separate elements, the points (or nodes), lines and polygons (or areas) described in Section 2.3. The database is topologically structured, that is, spatial relationships such as connectivity and adjacency are incorporated into the data structure. The DLG file also contains attribute or feature codes for the point, line and polygon elements. Areas (polygons) may have attribute codes which define the characteristic land cover of the area as, for example, forest, lake or swamp. Line features may be roads, railways, or rivers while point attribute codes include churches, telephone call boxes, or milestones.

Figure 2.7 shows a sample line graph. The corresponding digital records are given in Table 2.4. These records are not actual DLG records – they are given here to illustrate the concepts involved. The example shows 13 points, 14 lines and five areas. The points are each given a label (N1 to N13, the N referring to Node), and the $\{x,y\}$ coordinates of each point are stored. The polygons or areas are stored in a second file; again the individual areas are given identifying labels and a pair of coordinates defining an arbitrary point lying inside the polygon. Polygon 1 is the area lying outside the map and has an arbitrary centre $\{x,y\}$ coordinate of $\{0,0\}$. The lines which bound the polygons can be found from the Lines file, which shows the identifier of the polygon to the left and to the right. Area P2 is bounded by lines L1, L4, L5 and L14. These lines have entries of "2" in either the "left area" or "right area" column of the table. Notice that one point (N9) lies at the start and at the end of the same line, L12. That is possible only if L12 is a "degenerate line", that is, a line of zero length. N9 is a point that is not at the start or end of a real line segment and it has area 2 both to the right and to the left. The $\{x,y\}$ coordinates shown in the Lines file are the start and end coordinates of the line segment. In reality, a string of $\{x,y\}$ coordinates, up to 1500 pairs, would be used to define the position of the line.

The DLG format is a compact method of storing spatial data with a minimum of redundancy and with a structure which incorporates topological relationships.

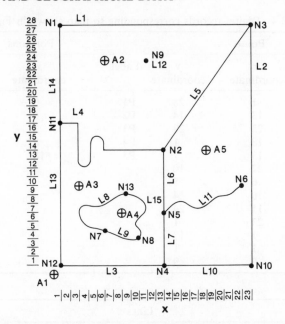

Figure 2.7   Sample line graph. (Reproduced by permission from *Data Users Guide 2*, Digital Line Graphs from 1:100 000-Scale Maps. US Geological Survey, Reston, Virginia, USA, Figure 2, 1985.)

Since {x,y} coordinates are stored using latitude and longitude the data can be plotted at any scale and on any appropriate projection. Details of US Geodata are obtainable from the National Cartographic Information Center, US Geological Survey, 507 National Center, Reston, Virginia 22092, USA.

2.5   SUMMARY

Geographical or spatial data may be measured on one of three scales – the nominal, ordinal and interval scales of measurement. Both the information content of the data and the cost of storage rise as the level of measurement is raised from the nominal to the ordinal and from the ordinal to the interval scale. The measurement scale chosen must reflect (a) the accuracy of the equipment used in the measurement process and (b) the use to which the data are to be put. Particular techniques of analysis are appropriate to each of the three scales; these techniques will be described in later chapters. Each data element is referenced to a spatial point, line or area; the line is defined in terms of a set of points while the area is defined by a set of lines.

Collection of geographical data is achieved using either manual or automatic means. A growing amount of spatial data is collected automatically by scanning devices on board orbiting Earth satellites, by digitizing instruments that can

Table 2.4 Digital records corresponding to map shown in Figure 2.7

	Points			Polygons	
Label	x coordinate	y coordinate	Label	x coordinate	y coordinate
N1	1	28	P1	0	0
N2	13	14	P2	6	24
N3	23	28	P3	3	10
N4	13	1	P4	8	7
N5	13	7	P5	18	14
N6	22	10			
N7	6	5			
N8	10	4			
N9	11	24			
N10	23	1			
N11	1	17			
N12	1	1			
N13	9	9			

	Lines					
	Points		Polygons		Coordinates	
Label	Starting	Ending	Left	Right	(First xy)	Last xy)
L1	1	3	1	2	1,28	23,28
L2	3	10	1	5	23,28	23, 1
L3	4	12	1	3	13, 1	1, 1
L4	11	2	1	3	1,17	13,14
L5	2	3	2	5	13,14	23,28
L6	2	5	5	3	13,14	13, 7
L7	5	4	5	3	13, 7	13, 1
L8	13	7	4	3	9, 9	6, 5
L9	7	8	4	3	6, 5	10, 4
L10	4	10	5	1	13, 7	23, 1
L11	5	6	5	5	13, 1	22,10
L12	9	9	2	2	11,24	11,24
L13	12	11	1	3	1, 1	1,17
L14	11	1	1	2	1,17	1,28
L15	8	13	4	3	10, 4	9, 9

record the x,y coordinates of points forming lines on maps (such as contour lines or coastlines) and by data loggers. Other data have to be digitized manually from maps or from field survey records.

Geographical data can be logically combined in the form of a Geographical Information System (Chapter 7) which is an integrated assemblage of hardware, software and data. The data can be envisaged as a set of transparent overlays, each overlay containing one item of data (for example, location of a particular type of crime, land use, rent level, or population details). Other ancillary data may take the form of tables of information relating to particular points or areas (such as details of plant species found in each field in a rural area or details of population age and economic structure). Such systems are now coming into operation; they demonstrate the validity of the statement that structured data are far more useful and thought-provoking than data which are recorded and stored in an amorphous or structureless form.

2.6 REVIEW QUESTIONS

1. Give definitions for the following terms:
   nominal, ordinal and interval measurement scales
   polygon   node   digitizer
   quadtree   vector   raster
   Geographical Information System
2. How do computer storage requirements relate to the measurement scale on which data are expressed?
3. Describe the advantages of a hierarchical data structure for raster data, such as the quadtree.
4. How can graphical data such as paper maps be translated into numerical or digital form?
5. Write a comparative assessment of data structures for vector data, indicating the relative advantages and disadvantages of each.

CHAPTER 3
Statistical Methods

3.1 INTRODUCTION

Many geography students seem to dislike statistics. The most probable reason for this is the perceived association between statistics and tedious calculation. Now that package programs such as SPSS, SYSTAT, SAS and BMD are widely available for both mainframes and PCs, the tedium of calculation can be avoided and more time devoted to the selection of suitable data and to the interpretation of results. Even so, a student will need to have some idea of the way in which the computational procedures which underlie statistical methods are carried out for, as Confucius wisely observed: "learning without thought is labour lost; thought without learning is perilous".

In this chapter some of the basic ideas which underlie statistical methods are introduced and examples given of their use in geography. A dataset of measurements on the 100 largest countries of the world in terms of population is used. This dataset, referred to in this chapter as the World Data Matrix, is described and listed in Appendix A. A copy of the dataset on an MS-DOS format disc is obtainable from the author.

The word *statistics* has at least three meanings. It can be used – as it frequently is in this chapter – to mean *techniques for summarizing and analysing data*. A second meaning, which is perhaps the original connotation, is *numerical data describing the economic and demographic characteristics of a state of region* (as in "economic statistics"). The third meaning is more specialized and less well-known; it is *one or more quantities derived from data by statistical analysis*. These quantities are used to test hypotheses about the properties of or relationships present in the data. An example used in this chapter is the statistic called "Student's t". The method of analysis is part of "statistics" as given in definition 1. The result of the test is a statistic (single number – definition 3) called t, while the data to which the test is applied may be statistics (definition 2). The meaning of the word statistics should be clear from the context, but it is necessary to be aware in advance of the different meanings that the word can take.

A distinction must be made between those statistical techniques which summarize data and those which are used to test hypotheses using observed data. The former are called *descriptive statistics* and the latter are *inferential statistics*. Provided that the data are expressed on an appropriate scale of measurement (as described in Chapter 2) then summary or descriptive statistics such as the median, the mean, and the standard deviation can be used to compress large sets of data into a smaller number of descriptive measurements (Section 3.2). The use of inferential statistical techniques requires the satisfaction of certain assumptions concerning the nature of the data, for each inferential statistical test has its own set of assumptions which must be met, even if only approximately, if the test is to be meaningful. The use of inferential statistics assumes knowledge of some basic terms, those of the probability distribution, the null hypothesis, the test statistic and the significance level. These terms are defined and illustrated in Section 3.4

3.2 DESCRIPTIVE STATISTICS

Descriptive statistics are used to summarize data. For example, few people are likely to want to know the temperature at noon at a given location for every day of the year; monthly averages are sufficient if one simply wants to compare the prevailing temperatures at two holiday resorts. Similarly, a single figure expressing the range or spread of values around the mean or average value provides a useful summary of the variability present in the measurement of a given variable. Two holiday resorts may have similar mean monthly temperatures but resort A may have warmer days and colder nights than resort B. This fact would be revealed if the spread of temperature values around the mean were to be expressed in terms of one number. The standard deviation of a set of measurements on a single variable serves this purpose. Neither the mean nor the standard deviation, when used in a descriptive way, involves anything more than data compression, whereas inferential statistics require that data conform to certain standards which are set out in more detail in Section 3.4

The World Data Matrix listed in Appendix A contains measurements on 14 variables for 100 countries. The variables are described in Appendix A, where a list of the names of the 100 countries will also be found. In those of the following examples which use the World Data Matrix it is assumed that an SPSS data file has been created using the method illustrated in Appendix B. SPSS (Statistical Package for the Social Sciences) is a widely-used set of statistical programs which is available for both mainframe and desktop computers. The term *variable* is used to mean *attribute or property* (the term is defined in more detail below). It is impossible to comprehend the information present in such a large amount of data such as that listed in Appendix A, so the first step in the study of the relationships among the 100 countries, such as: "which countries are most similar on the basis of these 14 variables?" or in the study of relationships between the variables, such as: "is the literacy rate greater in countries with a population less than 50 million than it is in those countries with a population greater than 50 million?", is to summarize the dataset in a concise way using *descriptive* statistics.

3.3 EXAMPLES: DESCRIPTIVE STATISTICS

3.3.1 Univariate summary statistics

The data contained in the World Data Matrix can be summarized using the SPSS CONDESCRIPTIVE command. This command is used to generate values of (i) the mean or average, (ii) the standard deviation (STD DEV in the listing below), (iii) the minimum and (iv) the maximum values in the data, together with a count of the number of valid data items in the dataset (VALID N). The SPSS commands for this operation are as follows, assuming that the instructions in Appendix B have been followed:

FILE HANDLE PMMSPSS / NAME = 'PMMLIB.PMMSPSS'
GET FILE PMMSPSS
CONDESCRIPTIVE ALL
FINISH

The FILE HANDLE command is described in Appendix B. The GET FILE command causes SPSS to read the dataset from file PMMSPSS into the memory of the computer, where it is operated upon by the CONDESCRIPTIVE command. The parameter ALL which follows the CONDESCRIPTIVE

Table 3.1   Output from the SPSS CONDESCRIPTIVE procedure for the 14 variables of the World Data Matrix listed in Appendix A

VARIABLE	MEAN	STD DEV	MINIMUM	MAXIMUM	VALID N	LABEL
POPN	49.176	135.913	3.80	1062.00	100	
AREA	1238.420	2905.137	1.00	22402.00	100	
POPCHNGE	1.986	1.157	− .20	3.90	100	
PCURB	47.110	25.455	5.00	95.00	100	
PCNONAG	57.650	28.310	7.00	99.00	100	
NATRES	3.050	1.766	.00	10.00	100	
ENERGY	1767.110	2294.713	3.00	9773.00	100	
STEEL	134.240	175.623	1.00	700.00	100	
FOOD	110.500	19.858	72.00	160.00	100	
TELE	14.610	22.656	.00	89.00	100	
NEWS	60.630	105.877	.00	441.00	100	
INFMORT	70.950	51.257	6.00	182.00	100	
LITERACY	62.710	31.569	3.00	100.00	100	
GNP	2992.300	4103.897	100.00	16400.00	100	

command means "include all variables in the dataset in the analysis". The output from this run of SPSS is shown in Table 3.1.

The column of Table 3.1 headed VARIABLE contains the names of the variables as defined in Appendix A. A more lengthy description of each variable could have been given in the definition of the SPSS data file (Appendix B), in which case that extended definition would have been printed under the column heading LABEL. Columns two to six contain the descriptive measures listed above. Using these descriptive measures it is possible to gain some insight into the range of values present in the World Data Matrix so that the values associated with a particular country can be interpreted relative to the mean or average value. Variable number 1, population in millions, and variable number 2, area in thousands of square kilometres, are included only for comparative purposes; they are not used explicitly in any of the analyses described below. Both population and area are difficult variables and it is usual to standardize the data so as to remove the effects of the magnitude of the population or the area, depending on the purpose of the analysis. Variables 3–14 in the World Data Matrix are expressed on a per capita or per cent/per thousand basis so as to make the values comparable.

Average values of the variables over the 100 countries show that, for example, the average percentage rate of population change in the 100 countries in the mid-1980s was 1.986% with a standard deviation of 1.157%. The standard deviation is a measure of the variability of the data. If the standard deviation is low then the data range is small; the converse is also true. The actual range of the data can be found by subtracting the minimum from the maximum value, giving a result of 4.1% for the population change variable. The country with the greatest percentage rate of population increase is Kenya, with a growth rate of 3.9%. Both Hungary and the Federal Republic of Germany had population growth rates of -0.2% in the mid-1980s, that is, the populations of these two countries were declining. These three countries were identified by finding the values 3.9 and -0.2 in column three of the World Data Matrix (Appendix A).

The mean, standard deviation, maximum and minimum of the other variables can be used in a similar fashion so as to get a feel for the data. It is important that a data user is familiar with the data being analysed, knowing, for example, the mean and the range of each variable, together with details of the scale of measurement of each variable. Definitions of the variables can be found in Appendix A.

Scrutiny of the minimum and maximum recorded values in the summary table can also help in the identification of errors resulting from mistyping or misreading the data. The definition of variable 6, for instance, shows that values outside the range 0–10 are impossible.

Descriptive statistics can be obtained for subsets of the data by the use of the SPSS "SELECT IF" command. In the next example, SELECT IF is used to subdivide the World Data Matrix into two parts, one consisting of countries with a population greater than or equal to 50 million in 1987 and the other consisting of countries with a population less than 50 million. The SPSS commands to perform the operation are:

```
FILE HANDLE PMMSPSS / NAME = 'PMMLIB.PMMSPSS'
GET FILE PMMSPSS
SELECT IF (POPN GE 50)
CONDESCRIPTIVE ALL
SELECT IF (POPN LT 50)
CONDESCRIPTIVE ALL
FINISH
```

These commands produce the output shown in Table 3.2 The upper table shows the summary statistics for the 21 countries which have a population of 50 million or more; these were picked out by the SELECT IF (POPN GE 50) command. In the command procedure above, the letters GE stand for "greater than or equal to". The lower table shows the summary statistics for the 79 countries with a population of less than 50 million, as the code LE, in the command procedure, means "less than". Comparison of the entries in column one of the two tables shows some interesting differences between the two groups of countries. The countries in the first group (population greater than or equal to 50 million) have a lower average population change than the countries in the second group (1.743% against 2.051%) and a higher proportion of their economically-active population working in areas other than agriculture (61.857% against 56.532%). Infant mortality in the countries of Group 1 is 61.571 per thousand live births; it is substantially higher (73.443 per thousand) in the countries of Group 2. Literacy rates are almost 10% higher in countries with populations of 50 million or more, and the range of variation in literacy (as shown by the standard deviation) is less in those same countries. Study of the variability of the values of the variables and of their ranges can also lead to the formulation of summary descriptions of the countries included in the World Data Matrix. The SELECT IF command can also be used to separate groups of countries on the basis of more than one variable.

3.3.2 Bivariate scatter diagrams

The preceding example was concerned with descriptions of the individual variables in the dataset and with comparison of the values of single variables for the two subsets. The use of scatter diagrams can assist in the comparison of the countries on the basis of two variables. The SPSS commands to draw a scatter diagram of the values of the 100 countries on two selected variables are:

Table 3.2   Output from the SPSS procedure CONDESCRIPTIVE after dividing the
            World Data Matrix into two sets using SELECT IF

NUMBER OF VALID OBSERVATIONS (LISTWISE) = 21.00

VARIABLE	MEAN	STD DEV	MINIMUM	MAXIMUM	VALID N	LABEL
POPN	180.652	260.880	50.40	1062.00	21	
AREA	3071.714	5373.986	143.00	22402.00	21	
POPCHNGE	1.743	1.078	− .20	3.20	21	
PCURB	49.667	24.906	13.00	90.00	21	
PCNONAG	61.857	25.462	28.00	97.00	21	
NATRES	2.571	1.287	1.00	5.00	21	
ENERGY	2115.333	2586.562	53.00	9577.00	21	
STEEL	169.810	201.047	2.00	569.00	21	
FOOD	115.619	17.007	83.00	141.00	21	
TELE	18.762	25.911	.00	76.00	21	
NEWS	73.762	114.808	.00	441.00	21	
INFMORT	61.571	44.147	6.00	140.00	21	
LITERACY	70.190	27.086	29.00	99.00	21	
GNP	3746.190	4813.693	150.00	16400.00	21	

NUMBER OF VALID OBSERVATIONS (LISTWISE) = 79.00

VARIABLE	MEAN	STD DEV	MINIMUM	MAXIMUM	VALID N	LABEL
POPN	14.227	10.187	3.80	46.00	79	
AREA	751.089	1470.242	1.00	9976.00	79	
POPCHNGE	2.051	1.175	− .20	3.90	79	
PCURB	46.430	25.712	5.00	95.00	79	
PCNONAG	56.532	29.069	7.00	99.00	79	
NATRES	3.177	1.859	.00	10.00	79	
ENERGY	1674.544	2219.597	3.00	9773.00	79	
STEEL	124.785	168.381	1.00	700.00	79	
FOOD	109.139	20.430	72.00	160.00	79	
TELE	13.506	21.759	.00	89.00	79	
NEWS	57.139	103.873	.00	422.00	79	
INFMORT	73.443	52.962	7.00	182.00	79	
LITERACY	60.772	32.523	3.00	100.00	79	
GNP	2791.899	3903.998	100.00	16380.00	79	

The upper table shows statistics for the 21 countries whose population is greater than or equal
to 50 million. The lower table gives statistics for the 79 countries with a population less than
50 million.

FILE HANDLE PMMSPSS / NAME = 'PMMLIB.PMMSPSS'
GET FILE PMMSPSS
PLOT TITLE = 'Infant mortality against population change'/
  PLOT INFMORT WITH POPCHNGE
PLOT TITLE = 'Infant mortality against percent urban population'/
  PLOT INFMORT WITH PCURB
FINISH

The FILE HANDLE command is explained in Appendix B. GET FILE is an
SPSS command to read the data from the file known to SPSS as PMMSPSS
and known to the Nottingham University ICL 3900 computer's VME operating
system as PMMLIB.PMMSPSS. PLOT TITLE gives a caption for the plot.
Note that, since the next command (PLOT INFMORT WITH POPCHNGE)
is an extension of the PLOT TITLE command, (a) the PLOT TITLE line
is terminated with a backslash character (/) and (b) the PLOT INFMORT
line starts in column 2. The command file given above produces two scatter
plots, one of infant mortality against the rate of population change and one
of infant mortality against percentage urban population. The result is shown
in Figure 3.1(a) and (b). The first scatter diagram shows that, in general terms,
infant mortality rates are lowest and increase very slowly where the percentage
increase in population is negative or small (up to around 2% per annum) whereas
there is little evidence of a simple relationship between the two variables when
the rate of population growth exceeds 2% per annum. Another interpretation
of Figure 3.1(a) is that infant mortality appears to increase linearly with annual
population change (that is, there is a proportionate increase in infant mortality
rates as the rate of population change becomes larger) but that the variability
of infant mortality rates also increases as the rate of population change increases.
  The second diagram (Figure 3.1(b)) shows a negative linear relationship
between infant mortality rates and the percentage of the population classified
as urban. The term "negative relationship" means that the line fitting the scatter
of points on a graph showing the relationship between the two variables slopes
downwards to the right, that is, as the percentage of urban population increases
so the rate of infant mortality decreases. Both scatter diagrams allow descriptive
generalizations to be made about the relationships between the two variables
depicted. Such relationships are called *bivariate*. The third and last example
in this section shows how all the variables of interest in the data matrix can
be used to describe the interrelationships between all the countries. This is an
example of a *multivariate* description.

3.3.3 Multivariate cluster analysis

The multivariate description of the World Data Matrix is an attempt to subdivide
the countries included in the data into sets in such a way that the countries in each

set are more similar to other countries in that same set than they are to any other of the countries in the other sets. Variables 1 and 2 (absolute population and area) are not included; the remaining 12 variables express the characteristics of the countries in terms of percentage, per thousand or per capita values. The method of cluster analysis is employed. It uses a measure of similarity between pairs of countries called the Euclidean distance. Each pair of countries is compared on the basis of the 12 variables using a measure which takes into account the differences in the values for each of the two countries on all the variables. For instance, the comparison between countries 1 and 2 is measured by the coefficient d_{12} given by:

$$d_{12} = (V_{11} - V_{12})^2 + (V_{21} - V_{22})^2 + \ldots + (V_{12,1} - V_{12,2})^2$$

where

d_{12} is a measure of the similarity of countries 1 and 2 on the basis of the 12 variables,

V_{11} is the value for variable 1 on country 1,

V_{12} is the value for variable 1 on country 2,

V_{21} is the value for variable 2 on country 1,

V_{22} is the value for variable 2 on country 2,

$V_{12,1}$ is the value for variable 12 on country 1, and

$V_{12,2}$ is the value for variable 12 on country 2.

The similarity measure d for countries 1 and 2 is therefore the sum of the squared differences in the values taken by countries 1 and 2 on the 12 variables in turn.

It is unreasonable to use this similarity measure directly on the raw data contained in the World Data Matrix simply because the different variables are measured in different units; the values taken by China and the USSR on variable 3 (population change) are 1.3% and 0.9%, a difference of 0.4%, whereas the same two countries have values of 55 and 550 on variable 8 (steel consumption in kg/head), a difference of 495 kg/head. The difference on variable 3 of 0.4 may seem to be insignificant in comparison with the difference of 495 on variable 8, yet the standard deviations of the two variables (1.157 and 175.623, see Table 3.1) are not the same. The standard deviation measures the variability in the measurements on a variable, so in terms of the overall world variation in population change the value 0.4 represents a proportion of the standard deviation of variable 3 of 0.38. The difference between China and the USSR on variable 8 is 495, which is equal to 2.82 times the standard deviation on that variable. The comparison should really be between the proportions of standard deviation (0.38 versus 2.82) rather than in terms of absolute differences (0.4 versus 495) otherwise variables will assume an importance proportional to the units in which they are measured. A transformation which converts from raw

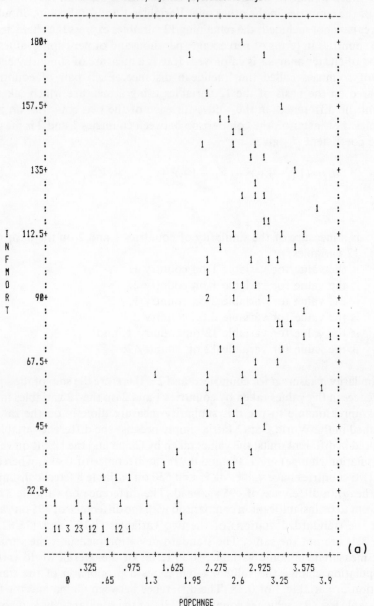

Figure 3.1   **(a)** Scatter diagram of infant mortality (deaths per 1000 live births) against percentage population change, using SPSS PLOT procedure. Note the elongation effect caused by the 6:10 aspect ratio of the standard lineprinter. **(b)** *(opposite)* Scatter diagram of infant mortality (deaths per 1000 live births) against percentage urban population, using SPSS PLOT procedure.

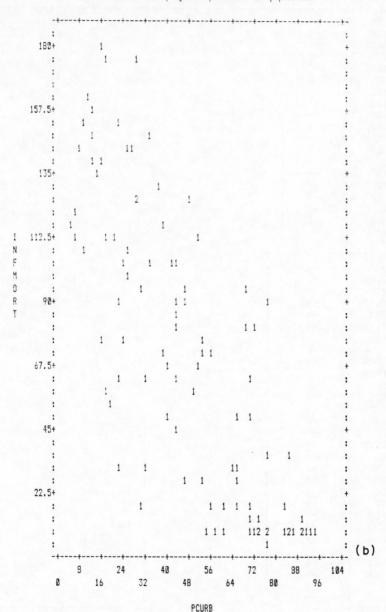

Infant mortality against percent urban population

(b)

PCURB

```
Dendrogram using Average Linkage (Between Groups)
                           Rescaled Distance Cluster Combine
  C A S E     0         5         10        15        20        25
 Label  Seq   +---------+---------+---------+---------+---------+
GUATEM  70    -+
HONDUR  93    -+
CAMERO  61    -+
ZAIRE   28    -+
ZAMBIA  77    -+-+
KENYA   38    -+ |
ZIMBAB  66    -+ |
TANZAN  34    -+ |
MADAGA  58    ---+-+
DOMINI  82    -+ | |
EL SALV 89    -+-+ |
PHILIP  13    -+ ` |
INDON    5    -+ | |
BURMA   25    -+-+ |
THAILND 18    -+   |
SRI LA  45    -+   |
CHINA    1    -+   |
MALAYS  47    -+-+ |
PARAGU  96    -+ | |
IRAN    21    -+ | |
ALGERIA 35    -+ +-+
IRAQ    43    -+-+ +-+
MOROCCO 32    -+ | | |
EGYPT   19    -+ | | |
TUNISIA 74    -+ | | |
TURKEY  20    -+-+ | |
IVORY C 57    -+ | | |
SYRIA   56    ---+ | |
PERU    40    -+   | |
ECUADOR 64    -+---+ |
BOLIVIA 85    -+     |
INDIA    2    -+     |
HAITI   87    -+     |
NIGERIA  8    -+     +------------------+
SENEGAL 78    -+     |                  |
BENIN   97    -+     |                  |
PAKISTN 10    -+     |                  |
NEPAL   42    -+     |                  |
BURUNDI 91    -+     |                  |
YEMEN   84    -+     |                  |
ETHIOP  22    -+-+   |                  |
NIGER   79    -+ |   |                  |
ANGOLA  71    -+ |   |                  |
SUDAN   33    -+ |   |                  |
BANGLA   9    -+ |   |                  |
KAMPUCH 83    -+ |   |                  |
MALAWI  75    -+ |   |                  |
GUINEA  86    -+ | | |                  |
CHAD    94    -+ +---+                  |
SOMALIA 72    -+ |                      |
SR LEON 99    -+ |         +------------------------+
```

Figure 3.2 Cluster diagram for the 100 countries in the World Data Matrix. The

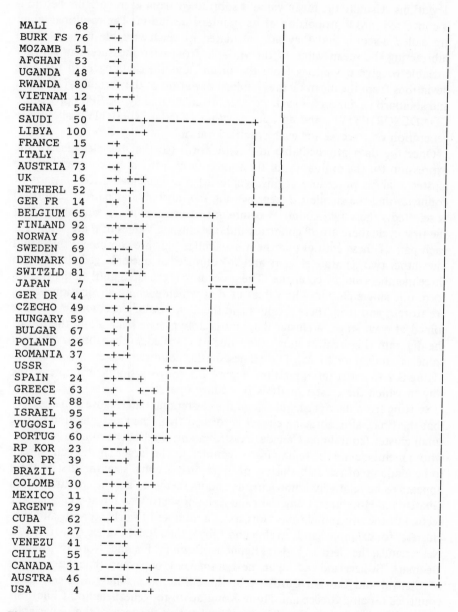

Figure 3.2 *(continued)* similarities between countries are calculated on the basis of variables 3–14 after standardization to z-scores. SPSS procedure CLUSTER was used to derive this diagram.

data to proportion of standard deviation evens out the units of measurement of the variables, for they are all being expressed in terms of their own standard deviations. Usually the mean value is subtracted from each variable before it is converted into a proportion of its standard deviation. The resulting values are called z-scores and they are calculated on each variable in turn by (i) subtracting the mean value of the variable from each measurement on that variable to give deviations from the mean, and (ii) dividing each of these deviations from the mean by the standard deviation of the variable. The mean and standard deviation for each variable are calculated by the SPSS procedure CONDESCRIPTIVE, and an option within CONDESCRIPTIVE allows the generation of z-scores for each specified variable.

Once the data are available in z-score form (so that the elements of the expression for the evaluation of the similarity measure d are comparable) the cluster analysis procedure begins. Values of d are calculated for all pairs of countries and the smallest d (representing the most similar pair of countries) is selected. These two countries combine to form a cluster, so at the end of the first cycle there are 98 countries and one cluster. Values of d are found for each pair of these entities and the most similar pair are combined; this could give either two separate clusters and 96 countries or one enlarged cluster with three members and 97 countries. The procedure is repeated until all the entities form one single cluster. The values of d at which each amalgamation occurs are stored, and from these d values and the identities of the countries that are joined at each stage, a cluster diagram is drawn. The scale along the top of the diagram (Figure 3.2) shows the value of d scaled so that the maximum d value is equivalent to 25. The names of the countries have been added to Figure 3.2 to assist interpretation. Figure 3.3 shows a simple example of the way in which the cluster analysis procedure operates.

Starting from the right-hand side of the cluster diagram (Figure 3.2) it is clear that the final, all-embracing cluster is formed from the amalgamation of one small cluster (containing Canada, Australia and the USA) and a large cluster whose members are the remaining 97 countries. The larger cluster can be seen to be made up of two sub-clusters of more-or-less equal size; the first of these appears to be relatively homogeneous and its members are the countries from Guatemala, Honduras, Cameroon and Zaire down to Vietnam and Ghana. The second of the two sub-clusters consists of a number of fairly separate sub-sub-clusters, for example Saudi Arabia and Libya, then Italy, France, Austria, UK, Netherlands, the German Federal Republic, Belgium, Finland, Norway, Sweden, Denmark, Switzerland and Japan; next, a smaller group comprising the countries of Central and Eastern Europe plus the USSR, and finally a larger set of countries (Spain, Greece and Hong Kong down to Venezuela and Chile). Six groups of countries can therefore be picked out on the basis of the 12 variables used. The cluster diagram is a multivariate description of the 100 countries, each characterized by 12 measurements. Further descriptive statistical

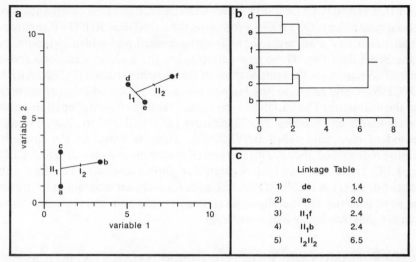

Figure 3.3 (a) Two variables define the axes on which points a–f are measured. The closest pair of points is d and e at 1.4 units distance. These points combine and are replaced by point I_1. Again, the closest pair of points is found; this time it is the pair a and c at 2.0 units distance. This pair of points is replaced by point II_1. The process is repeated until all points have been combined. The order of combination is shown in Figure 3.3(c). (b) Linkage tree showing the order in which pairs of points join using the information shown in Figure 3.3(c). Visual interpretation of this diagram reveals the presence of two well-separated clusters. One consists of points a, b and c and the second is made up of points d, e and f. Compare with the scatter diagram (Figure 3.3(a)). This type of analysis can be extended algebraically into more than three dimensions, although it is not easy to visualize a graph such as Figure 3.3(a) in more than three dimensions. (c) Tabular equivalent of Figure 3.3(b).

work could be undertaken by selecting each cluster in turn and using the SPSS procedure CONDESCRIPTIVE to produce the mean, standard deviation, minimum and maximum values for each group on the 12 variables used in the cluster analysis so that the reasons for the similarities within and differences between the groups could be revealed.

The SPSS commands required to carry out the cluster analysis are:

```
FILE HANDLE PMMSPSS / NAME = 'PMMLIB.PMMSPSS'
GET FILE PMMSPSS
CONDESCRIPTIVE ALL
    OPTIONS 3
CLUSTER ZPOPCHNG TO ZGNP /
    METHOD = BAVERAGE /
    MEASURE = SEUCLID /
    PLOT = DENDROGRAM
FINISH
```

The FILE HANDLE, GET FILE and CONDESCRIPTIVE commands have all been used before. OPTIONS 3 following the CONDESCRIPTIVE command requests that new standardized variables be created and added (temporarily) to the SPSS data file. These new variables are the z-scores described above. Their SPSS names are the same as those of the original variables (POPN, AREA, POPCHNGE and so on, as listed in Appendix A) with the addition of the letter Z at the beginning. The CLUSTER command requests a cluster analysis on the new standardized variables 3 to 12 inclusive (ZPOPCHNG to ZGNP) using a method of clustering called BAVERAGE which is based on the similarity measure d described above. This method of cluster analysis is called SEUCLID by SPSS. These are standard requests for cluster analysis. The final sub-command, PLOT = DENDROGRAM, asks for a cluster diagram to be drawn. The word dendrogram literally means a tree drawing and, as Figure 3.2 shows, a cluster diagram looks like a tree.

3.4   BASIC CONCEPTS FOR INFERENTIAL STATISTICS

3.4.1   Probability and the Gaussian distribution

Probability or chance is such a well-known concept that most people think that they know intuitively what it is. Unfortunately, intuitive concepts are not always well-understood. St. Augustine, writing about 1600 years ago, realized this when he wrote "What is time? If no one asks me, I know; yet if someone asks me, I do not know." For some people, the same may apply to the concept of probability.

To clarify matters, think of an event or occurrence which can be observed a large number of times on separate and independent occasions. An example might be the level of the water in a river. The water level is measured in centimetres above an arbitrary datum (zero) point. It is highly unlikely that the water level will be the same every time it is observed – it will change in one direction or another. Since river level fluctuates it can be said to be a variable quantity or simply a *variable*. In statistics a variable is an observable quantity which changes from time to time or from place to place. Temperature, barometric pressure, my bank balance and the value of the Stock Exchange index are examples of variables. Some quantities remain the same, for example the boiling-point of pure water at standard temperature and pressure. These latter quantities are called *constants*. The distinction between variables and constants is an important one.

Some variables change in value in a systematic fashion. They take on different values at different times, or in different places, but the way they change is fixed. For example, my age is a variable (it is changing continuously) but the change is systematic so that I can easily work out what my age will be on January 27th, 2009. A distinction has therefore to be made between variables which change in value in a systematic way, like my age, and those which change in value in an apparently random or unpredictable way, like the level of a river. The latter type are called *random variables* and they are the object of study of inferential statistics. In the case of my age (a systematic variable) I can work out or determine its value at any point in the future by simple arithmetic. A random variable can take on any of a very large number of possible values at any point in time and/or space and its value at any such point cannot generally be predicted accurately. For example, what will be the level of the Potomac River on January 27th, 2009? What will be the yield of wheat in North Dakota in the year 1996? These questions cannot be answered exactly because both variables (river level and wheat yield) are random variables. Because they are random variables, questions such as "what is the probability that the river level will be 3.1 metres above datum?" of "what is the probability that the wheat yield will be x tonnes per hectare?" are answerable by the use of inferential statistical methods provided that adequate and suitable data are available.

In some artificial examples probability values can be calculated from knowledge of the particular situation. If an unbiased dice is used the probability of throwing a two is 1/6, since all six outcomes are equally likely. In the long run, each of the six possible outcomes (1 to 6 inclusive) should occur an equal number of times since the probability (or chance) of each is one in six or 1/6. It may seem trivial to add that the sum of the probabilities for the six possible outcomes is $6 \times 1/6$ or one; this means that the probability of throwing a 1 or a 2 or a 3 or a 4 or a 5 or a 6 is 1 (that is, certain). The lowest possible probability value is 0, meaning that the outcome is impossible. The probability of throwing a 7 with a single throw of a normal dice is 0. Probabilities are, therefore, expressed as proportions on a 0–1 scale or, sometimes, as percentages on a 0–100% scale.

In most geographical problems it is not possible to calculate probabilities from prior information, as was done in the case of throwing an unbiased dice. The factors controlling the variable under study may not be known fully, if at all, and the effects of minor factors, together with the unpredictable nature of human behaviour, will be seen as an added random element. For example, even if all the relevant information about the population of the UK were available, it would still not be possible to calculate the probability that n people would visit the Asda supermarket in West Bridgford, Nottingham, on any particular day of the working week from this prior information. In reality, probabilities have to be estimated from a sample or selection of observations of events such as "person visiting Asda supermarket, West Bridgford". The topic of sampling is discussed below.

Rather than try to estimate a probability value for every possible value of the variable (as was done in the dice example) it is usual to divide the range of potential values of the variable into a number of classes. A probability value is then estimated for each class on the basis of observational data. In order to do this we need to know

- the range of the data, which is defined by the maximum and minimum values,
- the number of classes into which this range is to be subdivided, and
- the size or extent of each class.

Plots of the number of observations in each class (the class frequency) against class number can be carried out automatically by statistics packages such as SPSS. These plots are called histograms. The SPSS command is FREQUENCY with the sub-command HISTOGRAM. An example SPSS command file to produce a histogram of the 100 values of variable 14 (GNP) in the World Data Matrix is:

```
FILE HANDLE PMMSPSS / NAME = 'PMMLIB.PMMSPSS'
GET FILE PMMSPSS
FREQUENCIES VARIABLES = GNP /
   HISTOGRAM = NORMAL INCREMENT (500)
FINISH
```

The sub-command NORMAL requests that a dotted line representing the shape of the Gaussian probability distribution, explained below, be superimposed upon the histogram, while INCREMENT (500) means that the histogram classes should be 500 units wide (units are US dollars in the case of GNP). Evans (1983) gives some useful advice on the use of histograms and the choice of class intervals.

   Since the use of the HISTOGRAM option on the major statistical packages is relatively simple, the user can experiment with various settings of parameters such as class interval and number of classes until a satisfactory result is achieved. The histogram is a useful graphical way of presenting the independent measurement on a random variable. While the probabilities have intrinsic interest, it is their distribution, represented by the shape of the histogram, that is important to the present discussion because many techniques of statistical inference are based upon the assumption that the probabilities of the random variable or variables under study have a particular probability distribution, which is called the Normal distribution. This assumption is commonly known as the assumption of Normality. A more precise name is the Gaussian distribution, after the 19th century German mathematician, Gauss. The use of the term Gaussian is to be preferred, since the word "Normal" implies that the distribution is natural or expected. This is not the case; Gauss originally applied it to the distribution of errors of observation made by an astronomical telescope. Figure 3.4 shows that most of the observations of the values of a random variable whose probability distribution is Gaussian occur around the central or mean point, with the probability that a value x is observed diminishing in an exponential fashion as x becomes further from the mean point.

   Despite the fact that the Gaussian distribution was originally developed for use in studying the probability of error, it has become the standard model of

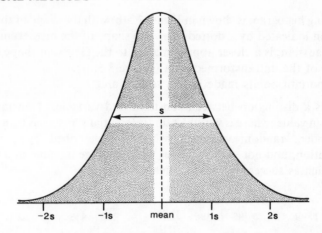

Figure 3.4 Gaussian or Normal distribution. The number of observations of a variable decreases symmetrically away from the mean, which is the centre of the distribution; in that sense the mean value is the "most likely". The rate of decrease in the numbers of observations away from the mean is measured by the standard deviation, s.

a probability distribution. Most of the best-known statistical tests are based upon the assumption that the probability distributions of the observations of the random variable under study are well approximated by the curve shown in Figure 3.4. Not all random variables have probability distributions that can be described by the Gaussian model. In such instances it may be possible to transform the observed values in such a way that the transformed values follow the Gaussian distribution more closely (Evans, 1983, pp. 139–148). For a set of observed values (x_i) with a few extremely high values, a square root or a logarithmic transformation is appropriate, that is, the values $\sqrt{(x_i)}$ or log (x_i) are used in place of x_i. The logarithmic transform is the more severe. Again, most of the readily-available statistics packages offer a range of data transformations. The SPSS command COMPUTE allows the generation of transformed variables; the following transformations can be used: ABS (value irrespective of sign), SQRT (square root), EXP (exponential), LG10 (logarithm to the base 10), and LN (logarithm base e). A sensible selection of such transformations could be combined with the use of the HISTOGRAM option in order to determine the transformation which produces the best approximation to a Gaussian distribution. An example SPSS command file to transform the variable GNP in the World Data Matrix to its base 10 logarithm and produce a histogram of the transformed values is:

```
FILE HAND PMMSPSS / NAME = 'PMMLIB.PMMSPSS'
GET FILE PMMSPSS
COMPUTE LNGNP = LG10(GNP)
FREQUENCIES VARIABLES = LNGNP / HISTOGRAM = NORMAL
```

The resulting histogram is shown in Figure 3.5(b) with the shape of the Gaussian distribution indicated by a dotted line. The shape of the histogram, although still not Gaussian, is a closer approximation to the Gaussian shape than is the histogram of the untransformed data (Figure 3.5(a)).

The important points made in this section are:

● There is a distinction between constants and variables. Constants do not change in value, whereas variables do. Inferential statistics is concerned with independent random variables, which are described by a probability distribution, and not with variables which change in value in a systematic way (such as successive measurements of age).

```
COUNT    MIDPOINT    ONE SYMBOL EQUALS APPROXIMATELY    .80 OCCURRENCES

  37      350.00    ****:*****************************************
  15      850.00    ****:***************
   8     1350.00    *****:****
   8     1850.00    *****:****
   3     2350.00    **** .
   1     2850.00    *    .
   3     3350.00    **** .
   1     3850.00    *    .
   1     4350.00    *    .
   2     4850.00    ***  .
   0     5350.00         .
   1     5850.00    *    .
   3     6350.00    ***:
   0     6850.00
   1     7350.00    * .
   0     7850.00      .
   2     8350.00    **:
   1     8850.00    *.
   3     9350.00    *:**
   0     9850.00    .
   0    10350.00    .
   3    10850.00    :***
   2    11350.00    :**
   1    11850.00    :
   0    12350.00
   0    12850.00
   0    13350.00
   2    13850.00    ***
   0    14350.00
   0    14850.00
   0    15350.00
   0    15850.00
   2    16350.00    ***
          I....+....I....+....I....+....I....+....I....+....I
          0        8       16       24       32       40
                    HISTOGRAM FREQUENCY
```

(a)

- Most of the standard statistical tests assume that the probability distribution of the observations of an independent random variable follows a standard form, called the Gaussian or Normal distribution.
- The probability distribution of a random variable can be estimated from its histogram. The probabilities for each class in the histogram are the class frequencies expressed on a 0–1 scale.
- A variable with a non-Gaussian probability distribution can be transformed to an approximate Gaussian form using, for example, square roots or logarithms.

3.4.2 Samples and populations

Recall the example above concerning the estimation of the probability that n people would visit the Asda supermarket in West Bridgford on a particular day of the working week, such as Saturday. It is not practicable or possible to measure the numbers of people visiting Asda on all Saturdays in the past or

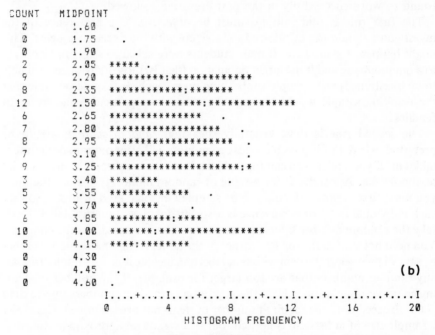

Figure 3.5(b) here

Figure 3.5 **(a)** *(opposite)* Histogram of Gross National Product (variable 14 in World Data Matrix, Appendix A) using SPSS procedure FREQUENCIES with the histogram class interval set to $500. **(b)** Histogram of Gross National Product (GNP) data after logarithmic transformation (base 10). The shape of the Gaussian distribution with the same mean and standard deviation as GNP is shown as a dotted line. The default number of classes (21) in the SPSS procedure FREQUENCIES is used. One symbol (*) equals approximately 0.4 occurrences.

in the future, so a *sample* is taken to represent the entire set of all possible observations. The entire set is called the statistical population. Inferential statistics is concerned with the problem of drawing reliable inferences about some characteristic of the statistical population from a sample drawn from that population. The use of a sample to represent or characterize a larger set of possible observations is crucial to the whole structure of statistical inference. No matter how good the statistical test might be, it will always produce an incorrect result if the sample is unrepresentative or biassed. For example, if a sample were to be drawn only from the inhabitants of San Juan Bautista, California, it might incorrectly be concluded that all the inhabitants of the United States speak Spanish. If a sample does not include a selection of all possible types, in proportion to their frequency, it will not be reasonable to infer the true nature of the whole set or population of observations. Since the proportions of each type or class are not usually known in advance, rules which have been found to work successfully in the past are often followed.

The first rule is that samples must be *objective*. The preference of the investigator should not be allowed to interfere with the selection process. This might happen, for instance, if male students were asked to select respondents to a questionnaire and if the target population (that is, those individuals available to be questioned) were people sunbathing on a beach. It is quite possible that the resulting sample would be biassed towards one particular age-group of females.

The second rule is that, generally speaking, *random samples* are to be preferred, where this is possible. Most computers will print out random number tables or, if you prefer, you can find tables of random numbers in many statistics textbooks (see Appendix D for a table of random numbers). To use the table you must first organize the data to be sampled in some kind of order, so that each individual event or occurrence is associated with a numerical label. Next, take the random number table and read across the rows or down the columns. You need not start at the top left corner of the table. If the first random number is, say, 96 then select the item with that label and include it in the sample. Discard any random numbers that are too large; for example, if the number of items in the target population is 6000 then ignore any random numbers bigger than 6000. Proceed until the required number of items has been sampled. Generally a sample size of at least 30 is needed, but this is only an approximate rule and guidance should be sought from a specialized text. For the present, it is sufficient to note that a sample size that is too small will give imprecise results (that is, estimates of population characteristics inferred from the sample will not necessarily be close to their true but unknown values) while a sample that is too large is a waste of time and energy and, perhaps, money.

Figures 3.6(a) to (d) illustrate the results of an experiment in which 2000 pseudo-random numbers were drawn from a Gaussian distribution with a theoretical mean of 500. The actual mean of the 2000 values was 499.2 and their standard deviation was 60.2. The frequency distribution of the 2000 values, which will now be considered to form a statistical population, is shown in Figure 3.6(a). Incidentally, this diagram is an example of a screen dump on a dot-matrix printer, as described in Section 4.2.2.2.

Experiment 1 entailed the independent derivation of 50 samples of size 5 from the statistical population of 2000 values. The mean was computed for each sample and the result is plotted in Figure 3.6(b) in which the true mean is shown by a solid line. It is clear that the 50 sample means fluctuate around the true mean but the degree of fluctuation or variability is high. This variability is measured by the *standard error*, defined as the standard deviation of the 50 sample means. The standard error for the 50 samples of size 5 is 31.83.

In experiments 2 and 3 the sample size was increased to 15 and 30 respectively, and the result is shown in Figures 3.6(c) and (d). The variability (as expressed by the standard error) reduces as the sample size increases; the standard error for Experiment 2 was 22.32 and for Experiment 3 it was 13.84. This is exactly what would be expected in the light of the guidelines given above, and shows that the means of the largest samples are more likely to be closer to the true mean of 499.2 than are the means of the smaller samples.

A good sample should be objective, random and of an appropriate size. It is used to estimate the true but unknown characteristics of interest of a population using techniques of statistical inference. A sample is needed because, in many cases, the number of items or events making up the statistical population is very large, or else the full population may not be available, so that it is neither feasible nor economical to measure the entire population. The use of samples that are statistically acceptable can lead to conclusions that have sufficient precision for the purposes of the exercise. For example, the population comprising all valley-side slope angles in Scotland is impossibly large, yet an estimate of the range within which 95% of all such slope angles are likely to lie can be obtained from a random sample of sufficient size. Estimates based on samples can be obtained relatively quickly and at such less cost than if the entire population were measured, even if that were possible.

By this stage the meaning of the terms *sample* and *population* should be clear and you should understand the purpose of inferential statistics, which is to draw

---

Figure 3.6 *(over page)* (a) Histogram showing the distribution of 2000 random numbers drawn from a Gaussian distribution. These numbers are considered to form the population on which the sampling experiments shown in Figure 3.6(b)–(d) were conducted. (b) Distribution of the values of the sample mean for $n = 5$. The true mean is shown by the horizontal line at $y = 499.2$. The spread of the sample mean values around the true mean is measured by the standard error. (c) As Figure 3.6(b) but with a sample size of 15. The reduction in the standard error is noticeable. (d) As Figure 3.6(b) but with a sample size of 30.

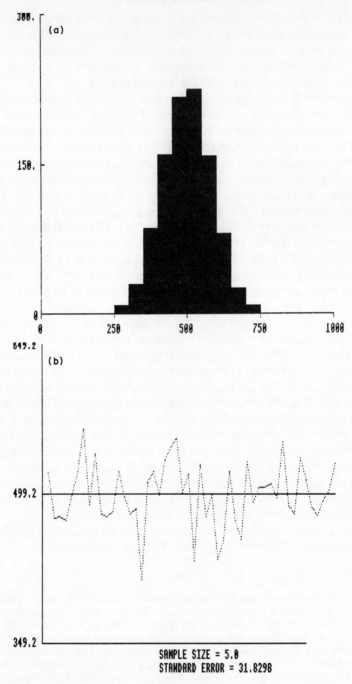

Figure 3.6

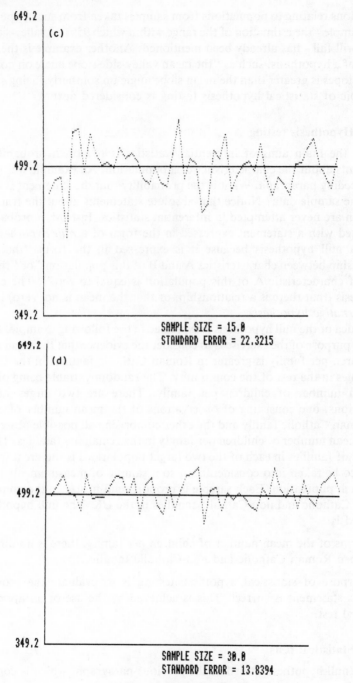

Figure 3.6 *(continued)*

conclusions relating to populations from samples taken from such populations. One example – the estimation of the range within which 95% of valley-side slope angles will fall – has already been mentioned. Another example is the formal testing of a hypothesis, such as "the mean valley-side slope angle on northerly-facing slopes is greater than the mean slope angle on southerly-facing slopes". The topic of statistical hypothesis testing is considered next.

3.4.3 Hypothesis testing

One of the main aims of inferential statistics is to attach probabilities to statements about the characteristics of a population. An example was given in the preceding paragraph. What is the probability that the statement is correct, given the sample data? Notice that absolute statements about the truth of an assertion are never attempted in inferential statistics. Instead, a probability is associated with a statement expressed in the form of a *null hypothesis*. It is called a null hypothesis because it is expressed in the form "there is no relationship between characteristics A and B of this population" or "the mean value of characteristic A of this population is equal to zero". The opposite hypothesis (that there is a relationship, or that the mean is not zero) is called the *alternative hypothesis*.

The idea of the null hypothesis is explained in the following example. Assume that the purpose of the exercise is to consider the evidence that the mean number of children per family is greater in Roman Catholic families in the UK than in families in the rest of the community. The random variable being observed is mean number of children per family. There are two target statistical populations, one consisting of observations of the mean number of children per Roman Catholic family and the other comprising all possible observations of the mean number of children per family in the remaining families. The total number of families in each of the two target populations is too great for every family to be taken into consideration, so a sample of measurements is taken from each population. Each sample is proportional in size to the numbers of Roman Catholic and non-Catholic families in the UK. The null hypothesis to be tested is:

> In terms of the mean number of children per family, there is no difference between Roman Catholic and non-Catholic families.

The purpose of statistical hypothesis-testing is to evaluate the probability that this statement is correct. This is achieved by the use of an appropriate statistical test.

3.4.4 Statistical tests

For the null hypothesis stated in the previous paragraph, which is concerned with the mean value of a characteristic measured for two samples, each drawn from a separate population, the test known as Student's t-test can be used. Student was the pseudonym of an early 20th century statistician – it does not refer to college students. Student's t-test, like other statistical tests, has various requirements or assumptions, chief among which is the Gaussian assumption which states that the probability distribution of the observations in each of the populations is Gaussian. For present purposes, it will be assumed that the data satisfy this requirement, though in practice one would study the histograms and perhaps apply transformations, as explained earlier. The *User Manual* of your statistics package will explain how to select the t-test and how to prepare and enter data for analysis. The output from the program will be the test statistic called Student's t. An example of the use of the SPSS T-TEST procedure is given later.

What does the value of Student's t mean? Before that question can be answered a hypothetical situation will be considered. Firstly, assume that the null hypothesis is, in fact, true and there is no difference in the mean number of children per family in the two communities under study. If the size of the sample of Roman Catholic families is N1 and the size of the sample of other families is N2 then it must be appreciated that the sample data represent only one out of a very large number of possible selections of N1 and N2 measurements from the two populations. The sampling procedure could be repeated a very large number of times, taking N1 measurements from the Roman Catholic population and N2 measurements from the remaining population each time. If each sample is random, (N1 + N2) different families will be chosen each time. The sample data will therefore be different, and so will the corresponding value of Student's t. A histogram based on the values of Student's t calculated for each set of N1 + N2 families could then be drawn up and a value found (say t_5) which was exceeded by only 5% of the values of Student's t derived from the many samples of data. It could be concluded that **if** the null hypothesis is true **then** the probability of getting a computed value of Student's t as high as, or higher than, t_5 is 0.05 or less (Figure 3.7).

Fortunately, repeated sampling is not necessary in order to discover the value of t_5. If the assumptions of the Student's t-test are satisfied then the values of t_5 for many combinations of N1 and N2 have been calculated and are available in *Tables of Critical Values*. A list of critical values of Student's t is contained in Appendix C. If the values of N1 and N2 were, for example, 35 and 50 respectively then, to use the table, add N1 and N2 and subtract 2 to give 83. The value 83 is called the number of degrees of freedom (the explanation of this term is not important to the present discussion). Find the value nearest to 83 in the left-hand column of the table in Appendix C (labelled "df" for degrees of freedom) and follow across the row until the column headed $p_2 = 0.05$ is reached. The value for 80 degrees of freedom (the value closest to 83) is 1.990 while the value for 90 degrees of freedom (the next highest value in the table) is 1.987. A value of 83 degrees of freedom can be interpolated – it is 1.989. This

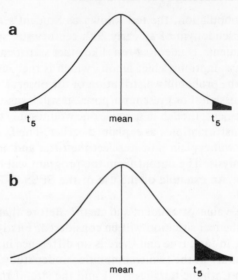

Figure 3.7   Distribution of values of Student's *t*. (a) The shaded areas represent 5% of the area under the curve. Thus, any value greater than the *absolute* value of t_5 has a probability of 5% or less of belonging to the population whose frequency distribution is given by the curve. Because both positive and negative tails of the distribution are used this is called a two-tailed test. (b) Only one tail of the distribution is used, so in this one-tailed test those values of *t* that are greater than t_5 are considered. A value of *t* greater than t_5 has a probability of 5% or less of belonging to the population whose frequency distribution is shown.

means: if a sample of size 35 is taken from population 1 and a sample from population 2 of size 50, AND if the null hypothesis is true, then the probability of obtaining a value of Student's *t* equal to or larger than 1.989 is 0.05 or less. If we had taken repeated random samples of size $N1 = 35$ Roman Catholic families and $N2 = 50$ non-Roman Catholic families and calculated the mean number of children per family for both groups, then – if the null hypothesis were true – the value of the Student's *t*-test statistic derived from these two means would equal or exceed 1.989 on 5% of occasions for a long series of such sampling tests.

It is not known, of course, that the null hypothesis is true. That is what one wants to find out. Consequently, if a value of Student's *t* greater than 1.989 is obtained, it could be concluded either that

- the pair of samples we have taken are unusual and the mean family sizes are truly identical, or
- the null hypothesis is not correct and there is a difference in the means of the two populations with respect to the random variable "mean number of children per family".

Which option would you choose? If you opt for the second alternative you stand a small chance (5%) of being wrong whereas if you opt for the first alternative you are saying that the samples really are quite unusual. Most people would prefer to take a 5% chance of being wrong and reject the null hypothesis in favour of the alternative hypothesis of unequal means. Notice that nothing is said about the magnitude of the inequality, and furthermore it is not specified whether the Roman Catholic families are larger or smaller than those of the rest of the community. If, as in this case, the direction of the difference is not specified in the alternative hypothesis then the test is called *two-tailed* (see Figure 3.7) and the column labelled p_2 in Appendix C is used. If the alternative hypothesis was that Roman Catholic families are larger than others then the direction of the difference would be specified and the test would be a *one-tailed* one, and the column labelled p_1 in Appendix C would be used. The distinction between one- and two-tailed tests is discussed further in Section 3.5.1.

The importance of the meaning of the terms *test statistic* (Student's t in this case), *significance level* (0.05 in the example) and *critical value* (1.989 in the example) should be clearly understood before other, possibly more complicated, statistical methods are used.

3.5   EXAMPLES OF INFERENTIAL TESTS

3.5.1   Differences in mean value

The World Data Matrix cannot be used directly in any inferential statistical test for one very simple reason: the population means and standard deviations for all the variables are known, since the 100 countries in the dataset constitute the entire statistical population. To use an inferential test of a self-evident hypothesis would be to turn logic on its head. It is possible to see by inspection of Table 3.2 that the mean gross national product of countries with a population greater than or equal to 50 million is greater than the mean GNP of countries with a population less than 50 million. Hence, there is no point in attempting to evaluate the probability of the truth of the statement that the two means are equal; inspection of the table shows that they are not.

For illustrative purposes only, and it is stressed that the following procedure is used solely to show how the various procedures work, a ploy will be used. Instead of using the entire World Data Matrix a random sample of 40% of the countries is taken and inferential statistical tests are employed to test various null hypotheses. Because the population means and standard deviations are available it is possible to check the veracity of the result of the inferential test. The method of sampling an available population is not a standard research method but is used here to simulate a real situation in which only a proportion of the population data is available through sampling.

The first example is one of a univariate or single-variable test. First of all, a 40% random sample is taken from the World Data Matrix. For three chosen variables in turn, the sample is subdivided into two groups, those countries with a per capita energy consumption of 1500 kg coal equivalent or more and those countries with an energy consumption of less than 1500 kg coal equivalent per head. If the means of the two groups for any one of the four variables gross national product, literacy, infant mortality and percentage of population classified as urban are denoted by X1 and X2 respectively, then the null hypothesis for each of the four separate tests can be written down as H0: X1 = X2. That is, on the basis of the sample evidence, there is no difference between the values of the means of the two groups and that the observed difference is due to chance. The Student's t-test, which was discussed earlier, is used to evaluate the probability of this null hypothesis given the assumption that the populations from which the samples are drawn each have a Gaussian probability distribution and that the variable concerned is measured on an interval scale.

The SPSS commands (i) randomly to sample the population of 100 countries at a rate of 40 per 100 (i.e. a 40% sample) and (ii) to carry out the t-test are:

```
FILE HANDLE PMMSPSS / NAME = 'PMMLIB.PMMSPSS'
GET FILE PMMSPSS
SAMPLE 40 FROM 100
T-TEST GROUPS = ENERGY(1500) /
    VARIABLES = GNP INFMORT LITERACY PCURB
FINISH
```

The FILE HANDLE and GET FILE commands should by now be familiar. SAMPLE 40 FROM 100 is the command to get a random sample of 40 cases (countries) from the 100 cases listed in the dataset. The T-TEST command invokes the SPSS procedure to carry out Student's t-test. GROUPS = ENERGY(1500) is the equivalent of SELECT IF; it tells SPSS to create two groups, one made up of those countries in the dataset which have an energy consumption of 1500 kg or more of coal equivalent per head and the other containing countries with an energy consumption of less than 1500 kg of coal equivalent per head. VARIABLES = introduces the list of variables on which a separate Student's t-test is to be carried out in turn. In the example, four independent t-tests are performed on the variables GNP, INFMORT, LITERACY and PCURB respectively. The variables do not have to appear on the VARIABLES = specification in the same order as they occur in the dataset.

The results are shown in Table 3.3. The means and standard deviations for each variable are printed separately for the two samples, together with the standard error. This is a measure of the precision of the sample estimate of the mean value, as explained above. If the population has a Gaussian probability distribution then the probability that the population mean of variable GNP for

Table 3.3 Output from the SPSS T-TEST procedure for four variables drawn from the World Data Matrix

T-TEST

GROUP 1 – ENERGY GE 1500.00
GROUP 2 – ENERGY LT 1500.00

VARIABLE	NUMBER OF CASES	MEAN	STANDARD DEVIATION	STANDARD ERROR	POOLED VARIANCE ESTIMATE			SEPARATE VARIANCE ESTIMATE		
					T VALUE	DEGREES OF FREEDOM	2-TAIL PROB.	T VALUE	DEGREES OF FREEDOM	2-TAIL PROB.
GNP										
GROUP 1	17	8000.5882	4907.035	1190.131	7.16	38	0.000	6.15	16.26	0.000
GROUP 2	23	653.4783	512.380	106.839						
INFMORT										
GROUP 1	17	28.4118	28.165	6.831	−6.37	38	0.000	−6.66	37.99	0.000
GROUP 2	23	97.9565	37.926	7.908						
LITERACY										
GROUP 1	17	85.2941	22.616	5.485	5.31	38	0.000	5.37	36.11	0.000
GROUP 2	23	44.9565	24.568	5.123						
POPCHNGE										
GROUP 1	17	1.1765	1.148	0.278	−4.99	38	0.000	−4.71	26.35	0.000
GROUP 2	23	2.6913	0.772	0.161						

A 40% sample of the data was taken and two groups of countries selected – Group 1 has an annual energy consumption of 1500 kg coal equivalent per head or more, Group 2 has an annual energy consumption of less than 1500 kg coal equivalent per head.

the Group 1 countries will be in the interval ($8000 + $1190) to ($8000 − $1190) is 0.682. Similarly, the sample mean for the Group 2 countries for variable GNP is $653 with a standard error of $106; given a Gaussian probability distribution, the probability that the population mean for GNP for the Group 2 countries lies in the range ($653 + $106) to ($653 − $106) is 0.682. In other words, the population mean GNP for Group 1 countries is estimated from the sample to lie in the range $6810 to $9190 with a probability of 0.682 while, with the same probability, the population mean for the GNP of Group 2 countries is estimated to lie in the range $546 to $760. The probable ranges containing the population means of the other three variables can be calculated in a similar fashion. However, while the range in which the population mean of each group might lie is relatively easy to calculate given the sample means and corresponding standard errors, it is less easy to make a judgement about the hypothesis that the population means for GNP (or one of the other variables) are in fact equal, with the implied conclusion that the difference between the calculated sample means of GNP is the result of chance in the random sampling process.

The Student's t-test allows the calculation of the probability that the two population means are equal. Before the Student's t-test is used, however, two questions should be answered. These are: (i) are we going to specify that the Group 1 mean is greater than (or less than) the Group 2 mean, or are we going to leave that question open? and (ii) is it possible to say whether the population standard deviations, that is, the standard deviations of GNP for the whole of the Group 1 countries and the whole of the Group 2 countries, are equal or unequal? Question (i) determines whether a "one-tailed" or a "two-tailed" test is used. A one-tailed test is one in which the investigator makes a prior statement that the alternative hypothesis H1 is of the form X1 > X2 (or X2 > X1, but not both) rather than simply stating as the alternative hypothesis that X1 is not equal to X2 (without specifying in advance which is the larger). If the latter choice is made then the test becomes a "two-tailed" one. The significance of this distinction will become clear in a moment. The answer to the second question determines whether (in SPSS terminology) the results in Table 3.3 under the heading "POOLED VARIANCE ESTIMATE" those under "SEPARATE VARIANCE ESTIMATE" are to be used.

It seems reasonable to assume *a priori* that the mean value of gross national product per head in countries with an energy consumption per head of 1500 kg coal equivalent or more is greater than the GNP per head of countries with a per caput energy consumption of less than 1500 kg coal equivalent. Consequently, the hypotheses to be used are (i) the null hypothesis H0: the statistical population means of the variable GNP/head of the two groups of countries are equal, and (ii) the alternative hypothesis H1: the mean GNP/head of Group 1 countries is greater than the mean GNP/head of Group 2 countries. A one-tailed test is therefore appropriate since the direction of the difference is specified in the alternative hypothesis.

It is also reasonable to suggest that, since the sample standard deviations of GNP/head for the two countries are so different ($4907 for Group 1 and only $512 for Group 2), the population standard deviations are also different. Hence the results in Table 3.3 under the heading "SEPARATE VARIANCE ESTIMATE" will be used. These results show that the value of the test statistic t is 6.15 and that the number of degrees of freedom is 16.26. The probability under the null hypothesis of getting a value of t as large as (or larger than) 6.15 is given in the third column, "2-TAIL PROB.". This probability is 0.000. But this probability refers to the two-tailed test. The probability appropriate to a one-tailed test is a half of the probability given for a two-tailed test. In this case it makes no difference, because a half of 0.0 is 0.0. The probability that the null hypothesis is true is therefore vanishingly small, and it would be reasonable to reject the null hypothesis and accept the alternative – on the basis of a 40% random sample it is highly probable that the mean GNP/head for countries with a per caput annual energy consumption of 1500 kg coal equivalent or more is greater than the mean GNP/head of countries with an energy consumption less than this value.

The results for variables INFMORT, LITERACY and POPCHNGE are interpreted in a similar way, except that the differences in the sample standard deviations of variables INFMORT and LITERACY are less than the difference noted above for GNP. It would not be illogical to use the Pooled Variance Estimate results for the middle two variables; in any case, because the sample standard deviations are similar the results for the Pooled and Separate Variance Estimates are reasonably close.

3.5.2 Relationships between variables

The second example of an inferential statistical procedure is concerned with explaining the variation in a variable of interest in terms of other variables. The variable of interest is called the *dependent* variable. Changes in the value of the dependent variable from country to country are assumed to be controlled by, or to be related to, corresponding changes in the values of each of a set of *explanatory* variables. The dependent variable might be difficult to measure and so it would be cheaper and easier to measure the explanatory variables and attempt to infer the most reasonable value of the dependent variable from the measured values of the explanatory variables. Alternatively, in developing an empirical model, one might wish to discover which member of a set of variables is most closely related to the dependent variable. The name of the statistical method used to answer these questions is *regression analysis*.

The example given in this section deals with a particular type of regression analysis in which there are several explanatory variables. These explanatory variables are labelled $X_1, X_2, \ldots, X_k$ and the dependent variable is Y. On the basis of a sample of n observations of $\{X_1, X_2, \ldots, X_k, Y\}$ the aim is to find the values of the terms $a_0, a_1, a_2, \ldots, a_k$ in the equation

$$\hat{Y} = a_0 + a_1X_1 + a_2X_2 + \ldots + a_kX_k$$

The terms $a_0, a_1, a_2, \ldots, a_k$ are the regression coefficients. The first coefficient, a_0, is the constant term. The coefficients are computed using a method called least-squares estimation which ensures that the sum of the squared differences between values of $\hat{Y}$ which are found from the equation above (the cap over the Y indicates that these are "predicted" values) and the observed values of Y is as small as possible (Figure 3.8(a)). In other words, if a regression analysis included two explanatory variables (X_1 and X_2) and the least-squares estimates of the coefficients a_0, a_1 and a_2 were 3.6, 2.1 and -9.6 then the multiple regression equation would be

$$\hat{Y} = 3.6 + 2.1X_1 - 9.6X_2$$

Assume that one of the sample observations was $X_1 = 3$, $X_2 = 5$ and $Y = -40$. Substituting the values of X_1 and X_2 into the equation gives the result $\hat{Y} = -38.1$. $\hat{Y}$ is the predicted value of Y. The difference between the predicted value (-38.1) and the observed value (-40) is the residual or error. In this case the error is ($-40 - (-38.1)$), the observed value of Y minus the predicted value of Y. The residual here is -1.9. The size of the residuals for each set of (X_1, X_2, Y) values in the sample gives an indication of the goodness of fit of the prediction. If the sum of the squared residuals (SSR) is small relative to the sum of the squares of the dependent variable about its mean then it can be concluded that the explanatory variables provide an adequate description of the variability in Y. The sum of the squares of the dependent variable about its mean is simply

$$SSY = \Sigma(Y_i - \overline{Y})^2$$

where the terms are defined as:

SSY   the sum of the squares of the dependent variable about its mean
Y_i the ith of n observations on the dependent variable
$\overline{Y}$ the mean value of the dependent variable
Σ means "the sum of the following expression".

The method of multiple regression used in this section is termed *stepwise*. This is because one explanatory variable is selected at each step. At the first cycle the single best explanatory variable X_a is selected. X_a is that variable out of all the available Xs which has the highest correlation with Y with the effects of the unincluded Xs being held constant. For example, if there were three explanatory variables (Xs) then the correlation between X_1 and Y with the effects of X_2 and X_3 held constant would be calculated, and the same would be done for X_2 and Y (with the effects of X_1 and X_3 held constant) and finally

X_3 and Y (with the effects of X_1 and X_2 held constant). These correlations are known as *partial correlations*. The topic of partial correlation is discussed further in the next paragraph. The explanatory variable with the highest absolute partial correlation with Y is selected in the first cycle. This explanatory variable is indicated by X_a. At the second cycle the explanatory variables considered do not include X_a, which was selected at cycle 1, and the explanatory variable with the highest partial correlation with Y is added to the regression equation and removed from the set of unincluded explanatory variables. This procedure is repeated until all Xs are included or until none of the unincluded Xs has a sufficiently high partial correlation with Y to justify its addition to the equation. The Xs are therefore added to the regression equation in order of decreasing partial correlation with Y, so that the sequence of regression equations would look like this:

(i) $\hat{Y} = a_0 + a_1 X_a$
(ii) $\hat{Y} = a_0 + a_1 X_a + a_2 X_b$
(iii) $\hat{Y} = a_0 + a_1 X_a + a_2 X_b + a_3 X_c$

and so on, until no further Xs are worth including.

The idea of partial correlation was introduced in the previous paragraph. The term correlation means related with, so it is easy to think of two variables that are correlated, such as a child's height and weight. However, both height and weight are correlated with a third variable, age, and as the child grows older its height and its weight usually increase. The correlation between height and weight is influenced by the fact that both variables are correlated with age. If the effects or influence of age can be removed then a clearer indication of the relationship between the height and weight of children will emerge. This is what partial correlation does – it removes the effects of other variables. In geographical studies area and population size are often correlated with many variables, so that the correlation between other variables is partly obscured by the fact that they are correlated with area and population size. That is why variables 1 and 2 of the World Data Matrix were not included in any of the previous statistical analyses.

As an example of the use of stepwise multiple regression, variable 12 (infant mortality per 1000 live births) is taken as the dependent variable and the candidate explanatory variables are percentage urban population, literacy rate, GNP per head, rate of population change, food intake per head, and energy consumption. The dependent variables are correlated, indicating that some information is being repeated, so it might be expected that not all of them will be chosen. The SPSS commands to perform the stepwise regression operation are:

```
FILE HANDLE PMMSPSS / NAME = 'PMMLIB.PMMSPSS'
GET FILE PMMSPSS
SAMPLE 40 FROM 100
REGRESSION
    VARIABLES      = INFMORT PCURB LITERACY GNP POPCHNGE
                     FOOD ENERGY /
    STATISTICS     = DEFAULT /
    DEPENDENT      = INFMORT / STEPWISE /
    RESIDUALS      = DEFAULT /
    SCATTERPLOT = (*RES,*PRE)
FINISH
```

Notice that a sample of 40 countries is taken randomly from the World Data Matrix, as explained above in the Student's t-test procedure. The REGRESSION command is self-evident, as is the STATISTICS command, which specifies the default set of statistical results. The DEPENDENT command defines the name of the dependent variable (INFMORT) and the method to be used (STEPWISE). RESIDUALS = DEFAULT asks for selected statistics concerning the residuals from the regression, and SCATTERPLOT (*RES,*PRE) requests a two-dimensional plot of residuals versus predicted Y values.

Table 3.4 lists the output from this SPSS run. The explanatory variable entered first (X_a in the terminology used earlier) is LITERACY. The "Multiple R" of 0.886 is the correlation between Y and X_a, while R Square is the square of this value. In the example, R Square is 0.78566; this is interpreted to mean that 78.566% of the variability in infant mortality can be explained by its relationship with literacy alone. This is a rather meaningless statement (but one which, nevertheless, is found in many statistics texts). It should be interpreted as follows: the $\hat{Y}$ values predicted from the regression of Y on X_a have a variation equal to 78.566% of the variation in the raw Y values. Just over 21% of the variation in the raw Y values is not accounted for by the relationship between Y and X_a. The next piece of interesting information occurs a little lower down the printout; it is an F-ratio value of 139.29. This F-ratio is a test statistic, and the null hypothesis is H0: if the whole statistical population had been used then the regression coefficient a_1 would be zero; it differs from zero in this instance only because a random sample has been used. The significance level for F is printed as 0.000 so the null hypothesis is not accepted as it has a virtually zero probability of being true.

The regression equation can be found from the values printed in the section of the table headed Variables in the Equation. The values entered in the column marked B are the coefficients of the equation, which in this case is: INFMORT = 154.59 − 1.38 * LITERACY. The symbol * indicates multiplication. The negative coefficient of − 1.38 means that as LITERACY rises by 1 unit, INFMORT falls by 1.38 units. The units of LITERACY are percentage points, while INFMORT is measured in units per 1000 live births.

At step 2 the variable PCURB is added to the equation. This is because its partial correlation with INFMORT (column 2, headed Partial, in the table Variables not in the Equation) is -0.471, which is the highest absolute value in the column. The Multiple R (the correlation between INFMORT on the one hand and LITERACY and PCURB on the other) is 0.912, compared with 0.886 for LITERACY alone. The R Square rises to 83.319% from 78.556%. The null hypothesis to be tested by the F-ratio is that the regression coefficients for both the included explanatory variables are zero. The value of F is 92.41 and the probability of getting a value of F as high as this if the null hypothesis were true is 0.000. The regression equation now becomes INFMORT $= 163.87 - 1.12 *$ LITERACY $- 0.52 *$ PCURB. Again, both regression coefficients are negative, meaning that as LITERACY and PCURB rise so INFMORT falls. The coefficient for LITERACY at this cycle is not the same as at cycle 1 because the regression relationship with INFMORT has been influenced by the inclusion of PCURB.

At cycle 3 the explanatory variable POPCHNGE enters with a partial correlation of 0.36. Multiple R and R Square rise to 0.924 and 85.440% respectively, rather smaller rises than those which occurred between steps 1 and 2. The null hypothesis that the regression coefficients in the full population are zero is again rejected (the F-ratio is 70.416 and the probability of getting an F-ratio as large as this if the null hypothesis were true is 0.000). The regression equation at step 3 is: INFMORT $= 129.83 - 0.91 *$ LITERACY $- 0.45 *$ PCURB $+ 8.64 *$ POPCHNGE showing that INFMORT goes down as LITERACY and PCURB go up but that INFMORT rises by 8.64 per thousand as POPCHNGE increases by 1%.

No further variables are added to the regression. This is signalled by the message PIN $= 0.050$ Limits reached, meaning that none of the remaining variables is important enough to warrant inclusion. The test of the importance of an explanatory variable is another F-ratio test, this time of the null hypothesis that the regression coefficient of the next candidate for inclusion in the regression equation is zero. If the null hypothesis is accepted it means that the next explanatory variable to be included in the regression equation has a regression coefficient which is not distinguishable from zero (it is non-zero in the sample only because of sampling error). In other words, the next explanatory variable would add nothing to the regression equation in terms of predicting the dependent variable, Y. Therefore, GNP, FOOD and ENERGY are left out of the final equation as the null hypothesis that the associated population regression coefficient is zero is accepted in each case.

Each regression coefficient in the final equation has an associated standard error. The term standard error was explained earlier in connection with the Student's t-test example. The population regression coefficient should lie within (plus or minus) one standard error of the sample regression coefficient with a probability of 0.688. It is good practice to list the standard errors of

Table 3.4   Results of stepwise multiple regression using SPSS procedure REGRESSION.
See text for discussion

Equation Number 1   Dependent Variable..   INFMORT

Beginning Block Number 1. Method: Stepwise

Variable(s) Entered on Step Number
1..   LITERACY

Multiple R	.88638
R Square	.78566
Adjusted R square	.78002
Standard Error	22.73192

Analysis of Variance

	DF	Sum of Squares	Mean Square
Regression	1	71977.47660	71977.47660
Residual	38	19636.12340	516.74009

F =   139.29145   Signif F =   .0000

Variables in the Equation

Variable	B	SE B	Beta	T	Sig T
LITERACY	− 1.387876	.117595	− .886377	− 11.802	.0000
(Constant)	154.587108	8.139232		18.993	.0000

Variables not in the Equation

Variable	Beta In	Partial	Min Toler	T	Sig T
PCURB	− .278983	− .470898	.610651	− 3.247	.0025
GNP	− .242128	− .417436	.637070	− 2.794	.0082
POPCHNGE	.267116	.397998	.475835	2.639	.0121
FOOD	− .205000	− .345908	.610248	− 2.243	.0310
ENERGY	− .195547	− .338096	.640730	− 2.185	.0353

Variable(s) Entered on Step Number
2..   PCURB

Multiple R	.91279
R Square	.83319
Adjusted R Square	.82418
Standard Error	20.32301

Analysis of Variance

	DF	Sum of Squares	Mean Square
Regression	2	76331.68913	38165.84456
Residual	37	15281.91087	413.02462

F =   92.40574   Signif F =   .0000

Variables in the Equation

Variable	B	SE B	Beta	T	Sig T
LITERACY	−1.115305	.134538	−.712297	−8.290	.0000
PCURB	−.521629	.160655	−.278983	−3.247	.0025
(Constant)	163.872305	7.818478		20.960	.0000

Variables not in the Equation

Variable	Beta In	Partial	Min Toler	T	Sig T
GNP	−.150892	−.264527	.491398	−1.646	.1085
POPCHNGE	.214994	.356550	.404086	2.290	.0280
FOOD	−.054473	−.082347	.381205	−.496	.6231
ENERGY	−.084925	−.146271	.471616	−.887	.3809

Variable(s) Entered on Step Number
  3..   POPCHNGE

Multiple R	.92434
R Square	.85440
Adjusted R Square	.84226
Standard Error	19.24921

Analysis of Variance

	DF	Sum of Squares	Mean Square
Regression	3	78274.44948	26091.48316
Residual	36	13339.15052	370.53196

F =   70.41628   Signif F =   .0000

Variables in the Equation

Variable	B	SE B	Beta	T	Sig T
LITERACY	−.906685	.156649	−.579060	−5.788	.0000
PCURB	−.454457	.154969	−.243058	−2.933	.0058
POPCHNGE	8.648116	3.776806	.214994	2.290	.0280
(Constant)	129.834563	16.607434		7.818	.0000

Variables not in the Equation

Variable	Beta In	Partial	Min Toler	T	Sig T
GNP	−.075187	−.126299	.367671	−.753	.4564
FOOD	−.028230	−.045387	.376355	−.269	.7897
ENERGY	−.025136	−.044208	.400386	−.262	.7950

End Block Number   1   PIN =   .050 Limits reached.

0 Outliers found. No casewise plot produced.

(continued)

Table 3.4   *(continued)*

Residuals Statistics:

	Min	Max	Mean	Std Dev	N
*PRED	− 3.1006	145.1678	68.4000	44.8000	40
*RESID	− 52.4176	50.5519	.0000	18.4940	40
*ZPRED	− 1.5960	1.7136	.0000	1.0000	40
*ZRESID	− 2.7231	2.6262	.0000	.9608	40
Total Cases =	40				

Outliers – Standardized Residual

Case #	*ZRESID
8	− 2.72310
25	2.62618
26	− 1.68672
24	− 1.58011
27	− 1.47299
21	1.35964
11	1.28580
30	− 1.17115
5	1.08216
40	1.01480

Histogram – Standardized Residual

```
NExp N        (* = 1 Cases,       . : = Normal Curve)
0   .03    Out
0   .06    3.00
1   .16    2.67 *
0   .36    2.33
0   .73    2.00 .
0  1.34    1.67 .
2  2.19    1.33 *:
4  3.23    1.00 **:*
5  4.25     .67 ***:*
3  5.01     .33 *** .
8  5.29     .00 ****:***
9  5.01    − .33 ****:****
3  4.25    − .67 ***.
0  3.23    − 1.00   .
2  2.19    − 1.33 *:
2  1.34    − 1.67 :*
0   .73    − 2.00 .
0   .36    − 2.33
1   .16    − 2.67 *
0   .06    − 3.00
0   .03    Out
```

Normal Probability (P-P) Plot

Standardized Residual

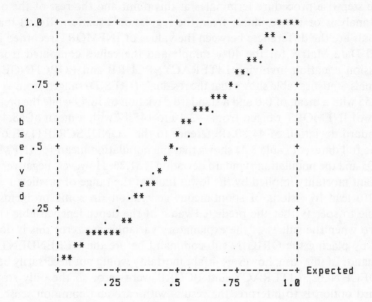

```
     1.0 +---------+---------+---------+------****
         !                                **  . !
         !                               *  .  !
         !                             *.     !
         !                           **.      !
     .75 +                          *.        +
         !                         *.         !
    O    !                       ***.         !
    b    !                     ***.           !
    s    !                   **  .            !
    e .5 +                  *.                +
    r    !                 *                  !
    v    !               .*                   !
    e    !             .**                    !
    d    !           .**                      !
    .25  +          .                         +
         !        . **                        !
         !      ******                        !
         !   **                               !
         !**                                  !
         +---------+---------+---------+---------+ Expected
         .25       .5        .75       1.0
```

Standardized Scatterplot

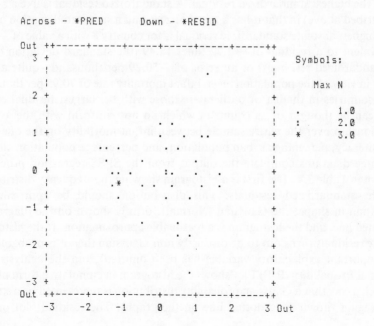

```
      Across - *PRED      Down - *RESID

   Out ++-----+-----+-----+-----+-----+-----++
     3 +                                     +   Symbols:
       !                                     !
       !                     .               !   Max N
     2 +                                     +
       !                                     !
       !                                     !   .   1.0
     1 +                 . ..                +   :   2.0
       !          .. . ..   .               !   *   3.0
       !        .  .  .  .  .     .          !
     0 +         ..  .  :. .   . .           +
       !         .*  . .   .                 !
       !           .         .               !
    -1 +              .                      +
       !                                     !
       !                                     !
    -2 +                  . .                +
       !                                     !
       !             .                       !
    -3 +                                     +
   Out ++-----+-----+-----+-----+-----+-----++
       -3    -2    -1     0     1     2     3 Out
```

regression coefficients so as to give some idea of the precision of estimation. The standard errors in the example are 16.61 (constant term), 0.16 (LITERACY), 0.15 (PCURB) and 3.78 (POPCHNGE).

The stepwise procedure terminates at this point and the rest of the output is an analysis of the residuals from the regression equation. Recall that the residuals are the differences between the values of INFMORT recorded in the World Data Matrix for the 40% sample and the values computed from the regression equation involving LITERACY, PCURB and POPCHNGE. The Residuals Statistics table shows that the residuals (*RESID) ranged from − 52.42 to 50.55 with a mean of 0.0 and a standard deviation of 18.49 while the predicted values of INFMORT ranged from − 3.10 to 145.17 with a mean of 68.40 and a standard deviation of 44.80. Reference to the CONDESCRIPTIVE output for the full dataset (Table 3.1) shows that the population mean for INFMORT is 70.95 and the population standard deviation is 51.26. However, negative values of infant mortality implied by the lower limit of the range of predicted values (− 3.10) lead to visions of spontaneous generation. In some problems it is sensible to specify that the predicted value of the dependent variable should be zero when the values of the explanatory variables are zero; this is done in SPSS by placing the ORIGIN sub-command before the DEPENDENT sub-command. In this case, however, infant mortality would not necessarily be zero if POPCHNGE, LITERACY and PCURB were zero, so the only straight-forward option is to interpret the results with care and common sense.

The next part of the output lists those cases (countries) in the sample which have the highest standardized residual. A standardized residual is like a z-score (described above) in that it has a mean of zero and a standard deviation of 1.0. The highest absolute standardized residual is for country 8 with a value of − 2.72, equivalent to a residual of − 2.72 times 18.49 (the standard deviation of the unstandardized residuals) or an error of − 50.29 per thousand – quite a large error in view of the population mean infant mortality rate of 70.95 per thousand. The countries in the list of outliers are those with the largest residual errors, and can be thought of as countries which do not conform with the average relationship over the entire sample between infant mortality on the one hand and literacy, percentage urban population, and percentage population change.

Three diagrams complete the output from the SPSS regression procedure shown in Table 3.4. The first is a histogram showing the frequency distribution of the standardized residuals. This distribution should be approximately Gaussian in shape; the Gaussian (Normal) form is shown on the diagram as a dotted line, and the histogram is a reasonable approximation. If the histogram of the residuals turns out to be distinctly non-Gaussian then it is probable that an important explanatory variable has been omitted from the analysis. The Normal Probability (P-P) Plot shows the histogram in cumulative form plotted in such a way that a Gaussian probability distribution is represented by a straight line (again shown by a dotted line on the graph). The residual plot on this graph is slightly S-shaped but does not depart very far from the ideal. Finally, the Standardized Scatterplot shows the predicted values of the dependent variable INFMORT plotted against the residuals from the regression. The plot should appear random with no evidence of a systematic trend. This is the case in the example, so it can be concluded that no significant explanatory variable has been omitted.

Regression analysis is a powerful way of testing statistical hypotheses. It can also be used to generate hypotheses concerning relationships present in a dataset. The stepwise procedure is used in the preceding example more as an exploratory tool ("What variables influence infant mortality rates most?") rather than as a means of confirming a preconceived hypothesis. Further details of the method can be found in Draper and Smith (1966), Edwards (1984) and Mather (1976).

3.5.3 Spatially-distributed variables: trend surface analysis

If the data under study consist of values of a single variable measured at a number of points over an area then one method of approaching the problem of describing and trying to account for the observed spatial pattern of variability in the values of the variable would be to set up a model describing that spatial variability or pattern. A commonly-used model considers the spatial pattern of values to be the sum of three separate components:

- an overall and smoothly-varying *regional trend*,
- *local deviations* from this trend, and
- a *random element*.

Usually, the investigation will be concerned with establishing the presence or absence of a regional trend, for instance in the annual rainfall amounts recorded at a series of measuring stations. If it exists, the trend might be correlated with some other spatial variable such as altitude. Local variations from the trend are useful in establishing a more detailed explanation for the levels of rainfall observed at particular locations within the region. Unfortunately, local deviations cannot be separated from the random element, which may be due to error of measurement or recording, or to interference with the measurement apparatus.

This example is concerned with the pattern of annual rainfall amounts over an area of Nottinghamshire. The purpose of the study is to discover whether the observed pattern shows any systematic, regional trend and, if so, whether it is possible to identify any factors which may be responsible for the trend. The null hypothesis is that the rainfall data show no systematic variation over the map area and any departure from a constant, uniform, value is the result of chance. This hypothesis implies that variability in recorded rainfall over Nottinghamshire is the result of a combination of random factors to do with measurement and sampling. The purpose of this study is to estimate the probability that the null hypothesis is correct.

To do this, data are required. In theory each of the infinite number of points lying inside the study area is available for sampling, but in practice the observations recorded at 41 rainfall stations are used as a sample drawn from this infinitely-large population. In order to test the null hypothesis a series of alternative hypotheses are presented and examined in turn, taking the simplest one first. The simplest alternative hypothesis states that the trend in the rainfall values is linear, that is, the true trend could be approximated by a plane surface. The position of this surface is fixed by the use of the principle of least squares which is illustrated in Figure 3.8. Once the position of this linear surface has been found from the sample of 41 points then the hypothesis of no trend (i.e. the surface departs from the horizontal only because of the choice of points in the sample of measurements) is tested. The test to be used was introduced in the multiple regression example above. It is named the F-test after the statistician Sir Ronald Fisher.

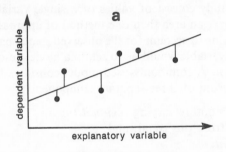

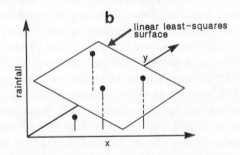

Figure 3.8 (a) A linear least-squares line relating a dependent variable (y) and an explanatory variable (x) is of the form $\hat{y} = a_0 + a_1 x$. The line is placed so that the sum of the squares of the differences between the measured and predicted values of y is the least possible for any straight line. These errors are often called *residuals*. (b) The same idea as that shown in Figure 3.8(a) can be extended to two explanatory variables. If these two explanatory variables are map x and y coordinates then the resulting least-squares surface is called a trend surface. This diagram shows a linear rainfall trend surface which relates rainfall over an area to map x and y coordinates.

Given a set of n measurements at points whose coordinates are given by x_i and y_i, the least-squares surface shown in Figure 3.8 can be estimated. The x_i and y_i are the map eastings and northings of the rainfall stations with the subscript i taking the successive values 1, 2, 3, . . ., 41. The rainfall at any point in the study area is estimated using a least-squares computer program from the sample of 41 points by the trend surface equation

$$\text{rainfall} = 686.58 - 24.73x + 14.61y$$

In this equation the terms x and y are the map coordinates of the point whose rainfall is to be estimated. The values 686.58, -24.73 and 14.61 are called the coefficients of the trend surface equation and they are found by the computer from the sample dataset. The first coefficient, 686.58, is called the constant term. It measures the mean value of the rainfall over the study area. The coefficient associated with the map x-coordinate (-24.73) shows the rate of change of rainfall in the x (easterly) direction; since the coefficient is negative it is apparent that rainfall is decreasing in an easterly direction. The coefficient associated with the map y-coordinate (14.61) indicates that rainfall increases in a northerly direction, but not as rapidly as in the westerly direction. Taking the x and y coefficients together it could be concluded that (on the basis of the sample measurements) rainfall in Nottinghamshire increases towards the west-north-west. If the equation is evaluated at each of a number of intersections of a regular rectangular grid over the study area a lineprinter can be used to produce a map of the linear trend surface for rainfall in Nottinghamshire (Figure 3.9). But is the surface shown a reliable estimate of the true trend surface (which could be known only if rainfall had been measured at every point in the area) or is it, as the null hypothesis states, an artifact resulting from random sampling of a flat, uniform rainfall surface?

This question cannot be answered with certainty. All that can be done is to estimate the probability of getting such a surface using 41 points if the null hypothesis were true. As in the preceding examples, a test statistic is needed. As indicated earlier, the test statistic used in this example is the F-ratio. Remember that the null hypothesis is: the sample of rainfall averages is drawn from a population which shows no spatial trend. This hypothesis implies that the true rainfall trend surface can be represented by a horizontal plane.

The value of the F-ratio as output by the computer program is 6.719. As with other test statistics, values called degrees of freedom are printed; these are used when looking up the critical value of the test statistic in tables. The F-ratio has two separate degrees of freedom, which in this case are 2 and 38. The instructions on the table of critical values of F tell us to use the smaller number of degrees of freedom (2) to determine which column of the table to use and the larger number (38) to determine the correct row. In column 2, row 40 (which is the nearest to row 38 printed in the table) the critical value of F at the 0.05 significance level is given as 2.44. Thus, if the null hypothesis were true then, if repeated,

```
MAP OF TREND SURFACE FOR ORDER 1.00

...........7777777777777............666666666666............$$$$$$$$$$$$$$$.
..........7777777777777............666666666666............$$$$$$$$$$$$$$$..
.........7777777777777............666666666666............$$$$$$$$$$$$$$$...
........7777777777777............666666666666............$$$$$$$$$$$$$$$....
.......7777777777777............666666666666............$$$$$$$$$$$$$$$.....
......7777777777777............666666666666............$$$$$$$$$$$$$$$......
.....7777777777777............666666666666............$$$$$$$$$$$$$$$.......
....7777777777777............666666666666............$$$$$$$$$$$$$$$........
...7777777777777............666666666666............$$$$$$$$$$$$$$$.........
..7777777777777............666666666666............$$$$$$$$$$$$$$$..........
.7777777777777............666666666666............$$$$$$$$$$$$$$$...........
7777777777777............666666666666............$$$$$$$$$$$$$$$............
777777777777............666666666666............$$$$$$$$$$$$$$$............5
77777777777............666666666666............$$$$$$$$$$$$$$$...........55
7777777777............666666666666............$$$$$$$$$$$$$$$...........555
777777777............666666666666............$$$$$$$$$$$$$$$..........5555
77777777............666666666666............$$$$$$$$$$$$$$$.........55555
7777777............666666666666............$$$$$$$$$$$$$$$........555555
777777............666666666666............$$$$$$$$$$$$$$$.......5555555
77777............666666666666............$$$$$$$$$$$$$$$......55555555
7777............666666666666............$$$$$$$$$$$$$$$.....555555555
777............666666666666............$$$$$$$$$$$$$$$....5555555555
77............666666666666............$$$$$$$$$$$$$$$...55555555555
...........666666666666............$$$$$$$$$$$$$$$..555555555555
..........666666666666............$$$$$$$$$$$$$$$.5555555555555
.........666666666666............$$$$$$$$$$$$$$$.5555555555555.
........666666666666............$$$$$$$$$$$$$$$..555555555555..
.......666666666666............$$$$$$$$$$$$$$$...55555555555...
......666666666666............$$$$$$$$$$$$$$$....5555555555....
.....666666666666............$$$$$$$$$$$$$$$.....555555555.....
....666666666666............$$$$$$$$$$$$$$$......55555555......
...666666666666............$$$$$$$$$$$$$$$.......5555555.......
..666666666666............$$$$$$$$$$$$$$$........555555........
666666666666............$$$$$$$$$$$$$$$.........55555555555555.
666666666666............$$$$$$$$$$$$$$$.........5555555555555.
666666666666............$$$$$$$$$$$$$$$...........555555555555.............4
6666666666............$$$$$$$$$$$$$$$.............5555555555555...........444
666666666............$$$$$$$$$$$$$$$$.............55555555555555..........4444
66666666............$$$$$$$$$$$$$$$..............5555555555555..........44444
6666666............$$$$$$$$$$$$$$$..............555555555555..........444444
66666............$$$$$$$$$$$$$$$..............5555555555555.........4444444
6666............$$$$$$$$$$$$$$$.............5555555555555........444444444
666..........$$$$$$$$$$$$$$$$.............55555555555555.......4444444444
66..........$$$$$$$$$$$$$$$.............5555555555555.........4444444444
6..........$$$$$$$$$$$$$$$.............5555555555555..........44444444444
..........$$$$$$$$$$$$$$$...............5555555555555.........444444444444
.........$$$$$$$$$$$$$$$.............5555555555555...........444444444444.
........$$$$$$$$$$$$$$$.............5555555555555...........4444444444444.
.......$$$$$$$$$$$$$$$.............5555555555555...........4444444444444..
......$$$$$$$$$$$$$$$.............5555555555555..........444444444444....
.....$$$$$$$$$$$$$$$.............5555555555555..........4444444444444.....
....$$$$$$$$$$$$$$$.............5555555555555..........4444444444444......

SURFACE ORDER   1.
TOTAL SS 65160.20
RESIDUAL SS 48098.91
EXPLAINED SS (%)26.18
```

Figure 3.9   Linear trend surface for Nottinghamshire rainfall data. The highest values are in the north-west and the lowest in the south-east.

different random samples of size 41 were taken, and if the rainfall were measured at each and the F-ratio computed for every sample, we should find that 5% of all the F-ratios exceeded 2.44. In other words, there is a 5% chance that F-value of 2.44 or greater will be produced when the null hypothesis is true. Our F-ratio is 6.719, which comfortably exceeds the critical value. The probability of the null hypothesis being true in the light of the sample evidence is much less than 5% and the alternative hypothesis that there is evidence of a linear spatial trend in the rainfall data is accepted. It has been shown that there is statistical evidence in favour of the hypothesis that the annual average rainfall in Nottinghamshire increases towards the west-north-west in a planar fashion, like the sloping roof of a house.

There may be a more complicated trend, for the River Trent flows across the study area and lower average rainfall might be expected within the valley than on the higher land to either side. To accommodate this hypothesis the surface could be allowed to develop a bend or a warp so that, while maintaining its WNW–ESE slope, it dips over the Trent valley. The trend surface that results from the adoption of this more complex hypothesis is shown in Figure 3.10; the bend representing the valley is apparent. But is this bend due entirely to random sampling effects? Could the pattern have arisen due to sampling of a population of rainfall measurements in which the true spatial pattern was represented by the planar surface? We need to test the null hypothesis that "the pattern exhibited by Figure 3.10 is due entirely to random sampling effects; the true regional pattern is a planar surface as shown in Figure 3.9". This null hypothesis is again tested using the F-ratio test statistic. Other, higher-order, surfaces with more complex patterns can be fitted and tested in a sequential manner. Details are provided in more advanced textbooks, such as Davis (1973) and Mather (1976).

3.6 SUMMARY

Some of the basic ideas of descriptive and inferential statistics are introduced and illustrated in this chapter. You should, by now, be familiar with the terms probability, sample, null hypothesis, significance level, and test statistic. Equally importantly you should be aware of the procedures of statistical hypothesis testing, which involve specification of a null hypothesis, selection of an appropriate statistical test, collection of a random sample of measurements, evaluation of the test statistic and comparison of the value of the statistic with the tabled value at a selected level of significance.

In the examples above, the test statistics used were Student's t and the F-ratio. There are, of course, many more and each is appropriate to a particular situation depending on the nature of the null hypothesis, the scale of measurement of the data and the probability distribution of the random variable being studied. A good introduction to the use of other statistical methods in

```
MAP OF TREND SURFACE FOR ORDER 2.00

..........................................999999.....8888.....7777.....6666....$$$$
..........................................999999.....88888....7777.....6666....$$$$..
.......................................999999......88888.....77777.....6666...$$$$$...
.....................................999999......888888.....77777.....6666....$$$$....5
..................................9999999......888888.....777777.....6666....$$$$.....55
...............................99999999......8888888......77777......66666...$$$$$....5555
............................99999999........8888888......777777.....66666....$$$$$.....5555.
.........999999999.........8888888.......777777......666666.....$$$$$....55555.
.999999999999..........88888888.......7777777......666666.....$$$$$.....55555...
999999...........88888888........7777777.......666666.....$$$$$$.....55555.....
...........8888888888.........77777777......666666......$$$$$$$.....55555.....4
....8888888888888.........777777777......6666666.......$$$$$$....55555.....44
88888888888..........7777777777........6666666.......$$$$$$......55555.....444
888.............77777777777.........66666666.......$$$$$$......555555.....4444
..........7777777777777777.........66666666.......$$$$$$$......555555.....44444
..777777777777777777.........6666666666.......$$$$$$$$......5555555.....444444
7777777777777............66666666666.........$$$$$$$$.......5555555.....444444.
7777.................66666666666.........$$$$$$$$$.......5555555......44444..
...................666666666666.........$$$$$$$$$$.......5555555......444444..
.........666666666666666666...........$$$$$$$$$$........55555555......444444...
6666666666666666666666.............$$$$$$$$$$$$.........55555555......444444...
6666666666666666666..............$$$$$$$$$$$$$.........55555555.......4444444...
666...............$$$$$$$$$$$$$.........55555555......4444444...
................$$$$$$$$$$$$$$$$........555555555......4444444...
........$$$$$$$$$$$$$$$$$$$$$$$........555555555......4444444...
$.......$$$$$$$$$$$$$$$$$$$$$$$$$$$.........555555555......4444444...
$$$$$$$$$$$$$$$$$$$$$$$$$$$$$$$$$$$$$.........555555555555......44444444..
$$$$$$$$$$$$$$$$$$$$$$$$$$$$$$$$$$$$$$$.........555555555555......44444444..
$$$$$$$$$$$$$$$$$$$$$$$$$$$$$$$.................555555555555.........44444444.
......$$$$$$$$$$$$$$$$$$$$$....................5555555555555.........444444444
..............................................5555555555555.........44444444
..............................................555555555555555........444444
..............................................5555555555555555..........44444
..............................................555555555555555555..........444
55...........................................5555555555555555....4
55555...............................555555555555555555.........
5555555.............................555555555555555555....
55555555.............................555555555555555555.....
555555555.............................55555555555555555....
555555555.............................55555555555555555
555555555....................$$$$$$......555555555555555555......555555
555555555................$$$$$$$$$$$$$$$$$$$$$$$$$.....................5
555555555.............$$$$$$$$$$$$$$$$$$$$$$$$$$$$$$$$$$$$$.........
55555555..........$$$$$$$$$$$$$$$$$$$$$$$$$$$$$$$$$$$$$$$$$$$$$$$$$$$$$......
5555555.............$$$$$$$$$$$$$$$$$$$$$$$$$$$$$$$$$$$$$$$$$$$$$$$$$$$$$$$$
5555555.........$$$$$$$$$$$$$$$.............................$$$$$$$$$$$$$$
555555.........$$$$$$$$$$$$$$$$...................................$$
55555.......$$$$$$$$$$$$$$$....................................
555........$$$$$$$$$$$$$$............666666666666666666666666666....
55.......$$$$$$$$$$$$.............6666666666666666666666666666666666
5........$$$$$$$$$$$..........6666666666666666666666666.................666666
SURFACE ORDER   2.
TOTAL SS 65160.20
RESIDUAL SS 31047.96
EXPLAINED SS (%)52.35
```

Figure 3.10   Quadratic (second-order) trend surface for Nottinghamshire rainfall data
Note the effect of the Trent valley, which is oriented in a SSW–ENE direction.

geography is Ebdon (1977), while Norusis (1983) and Frude (1987) give detailed examples of the use of statistical methods using SPSS. Other useful texts are Johnston (1978), Mather (1976), Norcliffe (1977) and Wrigley and Bennett (1981). Remember, though, that you will only become familiar with statistical methods if you use them.

The final point is more philosophical than practical. The availability of packages such as SPSS has led, in some instances, to the thoughtless use of statistical methods. You must have heard of the computer programmers' saying: "Garbage in, garbage out" or GIGO. Thought is, above all, vital. Statistical techniques enhance rather than replace thought. Data processing for its own sake is of no value; it must be done with some purpose, generally the testing of a well-framed hypothesis. Statistical packages have the advantage of relieving the user from the tedium of hand-calculation and allowing more time to develop hypotheses, collect sample data and think carefully about the meaning of your results. If your experience is typical, you will end with more unanswered questions than when you began; however, these new questions should be at a higher, more advanced level. Statistics help in developing geographical explanations of observational data by promoting and encouraging thought and reflection.

3.7   REVIEW QUESTIONS

1. Define the following terms:

Probability	sample	population
random variable	normal distribution	test statistic
null hypothesis	constant	histogram
trend surface	significance level	partial correlation
regression coefficient	standard error	

2. Why are samples used? What are the characteristics of a good sample?
3. What is meant by "rejecting the null hypothesis"? How can we be in error if we do reject the null hypothesis?
4. Define the terms "one-tailed test" and "two-tailed test". In what circumstances is each one relevant? Give appropriate geographical examples.
5. Describe how cluster analysis works. Using different shading patterns or colours and an outline map of the countries of the world, produce a map showing the geographical distribution of the different clusters of countries identified in Section 3.3.3. Comment on your results.
6. What is a regression coefficient? What does it measure?
7. Describe the model underlying trend surface analysis. Is it a realistic one?
8. Computers give the geographer more time to think about the results of statistical analysis. Has this been your experience?

CHAPTER 4

Computers in Cartography

4.1 INTRODUCTION

The end-product of the cartographic procedure is the map. Most people think of maps as hand-drawn paper-and-ink products, but now the production of maps by computer, either plotted on paper or drawn in the form of images on a graphics screen, is becoming commonplace. Digital maps are more easily adapted to a user's needs, especially when automated cartography is combined with spatial database management within the context of a Geographical Information System (Chapter 7).

Automated cartography, or digital mapping, is the process of storing, editing and generating maps using a computer. The production of block diagrams and other representations of spatial data in graphical form is also part of automated cartography. The effect of computer mapping techniques on traditional cartography has already been considerable. Digital mapping has been one of the strongest driving forces behind the development of Geographical Information Systems (GIS). Monmonier (1982, p. 2) believes that "the digital computer has had a profound effect on maps, an effect that will equal or surpass the changes in mapping occasioned by the invention of the printing press and the discovery of photography". Yoeli (1982) gives a very useful summary of many of the algorithms used in digitial mapping, including FORTRAN program listings.

Significant developments in the use of maps in the coming decades can be predicted with confidence. Signals from orbiting satellites forming the Global Positioning System (GPS) can be used to pinpoint the location of objects on the surface of the Earth to within a few centimetres; a computer mapping system carried within a car (and receiving the car's position from GPS) could, given suitable software, be capable of displaying a map showing the car's location and suggesting alternative routes, such as the shortest, fastest, most economical or most scenic route. Cooke (1987) reports that at least part of this scenario is now possible, for Chrysler demonstrated its Chrysler Laser Atlas Satellite System (CLASS), capable of storing 13 000 map images on optical disc, as long ago as the 1984 World's Fair in New Orleans.

Although the traditional concept of the map is changing, paper maps will continue to be widely used, because their use requires low technology (a pair of eyes, or even one eye), they are cheap to produce in large quantities, easy to store, and are well understood by the map-using community, although there is some evidence that many people are unable to relate the pattern on a map to the corresponding real-world features.

The major advantage of automated over manual cartography lies in the computer's ability to store cartographic and associated data and its speed in handling data and calculating results. The map shown in Figure 4.1 took only a few minutes to produce, and is easily amended and redrawn. A manually-drawn version of the same map would take many hours to produce. A manually-produced map cannot be easily changed, whereas a computer-drawn map takes its data from a database (as described in Section 2.4); this database can be readily edited and updated when new information becomes available, or amended when information becomes out of date, and a new map can then be automatically redrawn. In addition, because a computer-produced map can be redrawn quickly (or viewed on a graphics terminal) the cartographer can experiment with different contour intervals, or levels of shading, and can try different perspectives and projections with ease.

Maps can be generated digitally to suit different needs. For example, gas, water, and electricity utilities require maps showing the locations and attributes of pipelines or transmission lines, and local authorities need maps showing bus routes and frequencies, locations of street lights, or positions of police stations (to take a few examples). In the past, such maps have been time-consuming and expensive to make. Other users of digital maps will include automobile associations, who will keep base maps of their areas in digital form and quickly display a map showing the relative positions of the broken-down car and the nearest mobile service unit. Because the map coordinates of the locations of the car and the service unit are represented digitally, the shortest or most convenient route between the two points can be worked out automatically.

In this chapter the hardware and software aspects of automated cartography are described. Examples of the use of two widely-available software packages for automated cartography (SYMAP and GIMMS) are given, using datasets which are listed in the appendices. These datasets are also available from the author on a 5.25-inch floppy disc in MS-DOS format.

4.2 HARDWARE FOR COMPUTER MAPPING

The equipment (hardware) required for computer mapping can be divided into two kinds: that required to "capture" cartographic data from existing maps and that required to display newly-derived maps. Capturing data is called *digitizing*; it is the process of converting information to numerical form. Digitizing existing maps is essential if the data derived from decades of surveying

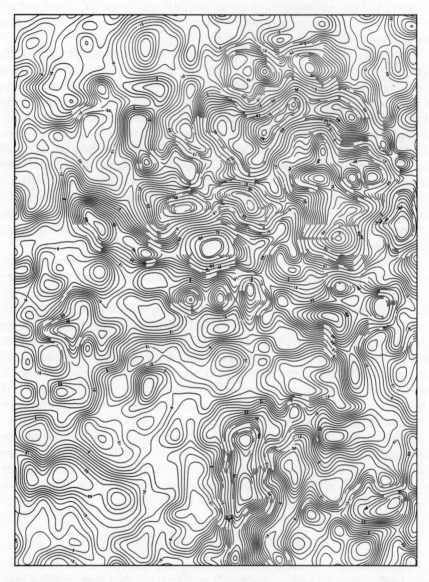

Figure 4.1   Machine-drawn contour map. (By courtesy of M. J. McCullagh.)

(topographic, geological, pedological, and so on) are to be made available to digital cartographic systems.

4.2.1 Hardware for digitizing

Digitizing means the conversion of information from analogue representation to numerical form. A paper map represents geographical information in analogue fashion, using lines and symbols. A digital map is essentially a numerical description of these lines and symbols. The (x,y) coordinates of points, lines (i.e. sets of points) and areas (defined by sets of intersecting lines), together with numerical descriptions or feature codes describing each point, line or area constitute the cartographic database. A digital mapping system requires a cartographic database in addition to the software and hardware required for the editing, compilation and production of maps and related products. A great deal of information already exists in the form of paper maps, and considerable efforts are presently being made to convert this analogue data to computer-compatible form. Digitizers are machines which perform this task.

A manual digitizer consists of a digitizing table and a pointer. The electronic hardware associated with the digitizer can determine the position of the pointer on the table whenever a button is pressed. Each map feature (whether it be a point or a sequence of points defining a line) is digitized placing the pointer at the location whose coordinates are required and pressing a button. The coordinates are then transmitted to the host computer. A description of the feature located at that point can also be entered into the host computer using a keyboard. The point may be a single entity, such as a church or other building on a 1:50 000 map, or it might be one of a string of points describing a linear feature such as a river, a contour line, a road, a railway or a boundary line. Each map entity (point, line or polygon) has a feature code or description associated with it in the database. The database itself is structured according to the uses to which the data are to be put (Section 2.4).

More elaborate digitizers can be used in "streaming" rather than "one-shot" mode. If the operator is following a linear feature, such as a contour line, the digitizer can be switched to streaming mode and the location of the pointer recorded at some specified time interval, such as every tenth of a second. The digitized points are therefore relatively far apart on straight, uncomplicated, sections of line but are more closely-spaced whenever a complicated curve is being followed.

Manual digitizing is slow and error-prone. If the map can be represented in raster format (Chapters 2 and 5) then image-processing operations can be used to enhance the map image so as to bring out linear features more clearly, and automatic line-following algorithms can be employed to generate the numerical coordinates of points along the lines. Whenever the algorithm cannot decide unambiguously which line to follow (for example, at the intersection of two roads, or at a railway junction or a river confluence) then the operator is asked to resolve the ambiguity by making a choice. No algorithm is yet capable of deciding on a feature code for the line being followed, so the operator is also required to add feature codes describing each line (for example as a motorway, a pylon line or a county boundary). The fineness of the detail that can be seen on the rasterized image depends on (i) the size of the pixels, termed the spatial resolution of the system, and (ii) the number of grey-levels that can be represented by the system. On a drum-scanning system the map to be digitized is laid on a rotating drum and a sensor is passed across the map; one pass of the sensor generates one line of the raster. At each of a large number of points along the raster the sensor records the level of grey in its area of view. These points are called pixels (for picture element). The grey level is expressed as a number, normally on a 0–255 or 0–1023 scale, so that the resulting raster representation of the map contains sufficient spatial and contrast resolution.

Cheaper rasterizing systems can be built around a conventional remote-sensing image display system such as that described in Chapter 5. A vidicon TV camera can provide an input to such systems, and the image from the TV camera is stored in a memory bank just as if it had been read from a disc file (as would be the case with a remotely-sensed image (Chapter 5)). The pixel size of the TV-scanned image is much larger than the pixel size generated by a drum scanner. TV-scanned images suffer from two problems in addition to those of low spatial resolution (i.e. large pixel size) and relatively low contrast. The two problems are (i) uneven illumination of the map and (ii) geometrical distortion introduced by the optical system. Where high-quality digitizing is not justified, for example in cases where a "picture" of a map is required simply as a background, then TV camera input is a cost-effective means of providing such a raster backdrop.

Details of the types of data collected by national mapping agencies are discussed briefly in Chapter 2. The Ordnance Survey of Great Britain, as the primary mapping agency for England, Scotland and Wales, is engaged in a digitizing programme which will result in complete digital map coverage of Great Britain by 2005. The Ordnance Survey distinguishes between basic scale maps (at 1:1250, 1:2500 and 1:10 000 scales) and derived scale maps (1:25 000, 1:50 000, 1:250 000 and 1:625 000). Some 1:10 000 maps are derived from basic scale 1:1250 and 1:2500 maps. Digitizing began in 1973 and completion of the digitizing of the 1:1250 basic scale maps is expected by 1993 and of the 1:2500 series by 2005. This work is accomplished by manually digitizing the field surveyors' Master Survey Drawing (MSD) which consists of the current published map sheet plus amendments noted by the surveyors. The Chorley Report (1987) notes that it costs around £800 to digitize, check and edit one Master Survey Drawing. The end product is a standard format magnetic tape which contains (i) coordinates defining points, lines and polygons, and (ii) feature codes for each line, point or polygon. The feature codes (of which there are 160) indicate the label to be given to the line, point or polygon such as fence, road, church and so on. A single digital map currently costs £85.

The Ordnance Survey of Great Britain is now investigating methods of accelerating its digitizing programme. In particular, rather than using manual digitization of the Master Survey Drawing, which involves a skilled operator following each line and noting the position of each feature with a pointer, as described above, the Ordnance Survey is considering the use of raster-based digitizing techniques.

The US Geological Survey's National Cartographic Information Center is also marketing digital cartographic data under the title of US GeoData. Digital map data are available for each 7.5- or 15-minute quadrangle in four data files as described in Section 2.4.2.4.

4.2.2 Hardware for map production and display

Digital maps and related products, such as block diagrams, can be displayed either as images on a graphics terminal or in hard-copy form. The most common form of hard-copy output device is the plotter, but for draft-quality maps and diagrams a standard line-printer or dot-matrix printer can be used.

4.2.2.1 Graphics display terminal

A graphics display terminal is a high-resolution TV set and, like a TV set, it can display either a monochrome or a colour image. The screen is composed of an array of phosphor dots (red, green and blue-sensitive triplets in the case of a colour display screen) and these dots are arranged in the form of a matrix. Low-resolution systems such as those associated with standard microcomputer displays have screen resolutions ranging from 320 (horizontal) × 200 (vertical) to 640 × 480 (the IBM CGA and VGA systems). More advanced systems have screen resolutions of 1024 × 1024 or greater. Higher screen resolutions mean either that greater detail can be perceived since the phosphor dots are closer together, or that a greater map area can be displayed at a larger scale. However, increasing the screen resolution means that more data (defining the dot patterns) must be sent to the screen. Since a TV screen works on the refresh principle, that is, the image on the screen is updated every 1/50 second, the data forming the displayed image must be sent from the computer to the screen every 1/50 second or the picture will fade. For a screen resolution of 1024 × 1024 dots and a bi-level image with 0 representing black and 1 representing white the number of bits (Chapter 1) to be transmitted each second is 52 428 800. If the resolution is increased to 2048 × 2048 then the number of bits per second to be transmitted rises to 209 715 200.

These values can be halved if the screen display is refreshed using the *interlacing* technique which updates each odd-numbered line of the raster every odd-numbered refresh cycle and updates the even-numbered rasters every even-numbered cycle. Each raster is thus refreshed every 1/25 second if interlacing is used, and this reduces the data transmission requirements to 26 214 400 bits per second (c. 26 MHz) and 104 857 600 bits per second (c. 105 MHz) respectively. TV monitors differ in their data-handling capabilities – a cheap monitor will not be capable of refreshing the screen at the rate required by a high-resolution output device. Another consideration is that monitors working in interlaced mode produce an image that generally is more prone to flicker than a non-interlaced display of the same screen resolution.

Colour display systems available for personal computers such as the IBM VGA allow the use of up to 16 colours (light and dark versions of the primary colours red, green and blue together with their combinations – yellow, cyan and magenta – plus black and white) at a resolution of 640×480, or 256 colours at a reduced screen resolution of 320×200. Because a colour system requires three dots (one for each primary) at each point on the screen, resolution tends not to be as good as a similarly-sized monochrome display. The number of colours available depends on the number of levels of the three primaries. If there are only two levels (on and off) available then the number of colours including black and white is 2^3 or 8. If eight levels of each colour can be shown then the total displayable colours number 2^8 or 256. Increasing the number of displayable colours or shades of grey requires more memory to store the image. More details of colour monitors are provided in Section 5.5.

4.2.2.2   Dot-matrix and inkjet printers

Both vector (line) maps and choropleth (area-shaded) maps can be produced on a printer. The quality of vector maps is substantially lower than that of maps produced on a specialized device such as a high-resolution screen (Section 4.2.2.1) or a pen-plotter (Section 4.2.2.3), but the dot-matrix printer is both cheap and widely-available. As the name implies, a dot-matrix printer uses patterns of dots printed within a rectangular matrix to generate individual characters. The Epson RX-80 printer is typical of the cheapest dot matrix printers on the market. It uses a vertical column of nine dots moving laterally across the page to produce printed characters, each of which is generated as a dot pattern within a matrix measuring six columns by nine rows. The letter T as printed by an RX-80 is shown in Figure 4.2. Each dot is produced by firing a pin against an inked ribbon.

In normal character-printing mode a program sends instructions to the printer to print a character string in a form such as PRINT "⟨message⟩" and the printer hardware selects the appropriate dot patterns for each of the characters in the character string "⟨message⟩". The dot patterns for all printable characters are stored internally by the printer, the codes corresponding to the normal ASCII character set described in Chapter 1. These dot patterns are based on a 9-row

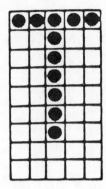

Figure 4.2   Letter T in the Epson RX-80 character font. Five vertical strikes of the print-head are needed to produce a character; each strike uses up to eight of the nine available pins. The ninth row is used for descenders (such as the tails of the letters g and y). A blank column separates adjacent characters.

by 5-column matrix on an Epson RX-80 printer; the sixth column is left blank to form a vertical space between characters. Other, more expensive, printers use a larger matrix; for example, the Mannesman-Tally MT-86 builds its characters in an 8×11 grid, while some printers use a 24-dot print head.

Choropleth maps are maps which show the distribution of the magnitude of a variable such as population density for each of a set of areal units such as counties or enumeration districts. The range of values taken on by the variable is divided into a number of classes before the map is drawn. If such maps are produced by hand or drawn by a pen plotter, the increasing magnitude of the variable is shown by increasing density of line-shading. Counties with low population density have sparse shading while counties with high population density have dark shading. If the map area is represented by a matrix of cells, then the shading level in each cell can be approximated by a printable character (the upper and lower case letters A, B, C, . . ., Z, the digits 0–9, and symbols such as *, +, ., and $) or combinations of these letters, digits and symbols. A dark shading level is produced on a printer by overlaying a combination of characters such as X, *, + and M. An example of a printed choropleth map is shown in Figure 4.9 (pp. 124–6). MacDougal (1976) treats the topic of line-printer map production in detail.

Program packages such as SYMAP use the principle of overprinting characters to represent different levels of shading on a printer. SYMAP can be used on line-printers as well as dot-matrix printers; the principle is the same, the only difference being that the line-printer outputs a full line at once rather than one character at a time, as is the case with the dot-matrix printer. Figure 4.9 shows a map produced on a standard line-printer by the SYMAP package. The advantages of this kind of map are cheapness and rapidity of production but these advantages have to be offset against the crudity of the printer map relative to the high-quality pen-plotter map. Nevertheless, many exploratory applications do not require high-quality output and the SYMAP type of produce is adequate.

(a)

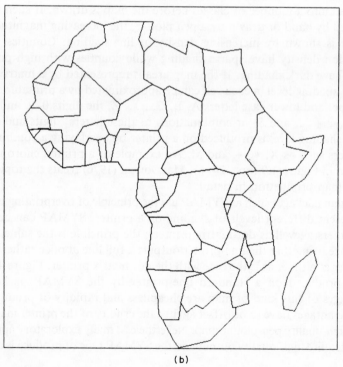

(b)

Whereas a line-printer can output only entire characters, the dot-matrix printer can output individual dots in the 6 by 9 dot matrix mentioned above. Thus, any desired pattern of dots can be sent to the printer. The printer is said to be in bit-image mode when this happens, for the code to fire a pin is the digit 1 and the code to leave a blank is the digit 0. In normal bit-image mode an 80-column printer can be programmed to print 540 dots horizontally across the paper, with the number of dots printed vertically being restricted only by the length of the paper. The number of dots per inch in the horizontal and vertical directions can be set to 72 to ensure equal vertical and horizontal scales. The instructions needed to set the printer into bit-image mode are called escape codes because each instruction is preceded by the ASCII code 27 which is generated by the ESC or ESCAPE key on a terminal keyboard. For example, the code ⟨ESC⟩ A 8 will set the spacing between lines on an Epson-compatible printer to 8/72 inch. In bit-image mode the ninth pin on the print head (Figure 4.2) is not usually used so eight vertical pins plus the gap between the bottom pin on line i and the top pin on line $i + 1$ will occupy 8/72 inch because the spacing between pins is 1/72 inch. Each byte that is printed in bit-image mode can be thought of as an eight-bit binary number with the digit 1 meaning "print a dot here" and the digit 0 meaning "do not print a dot here". Eight rows of dots are printed at once by a single horizontal move of the print head across the paper. If the paper is moved upwards by exactly 8/72 inch before the next line of eight-bit bytes is printed then there will be no gap between the lines.

Because the dots in the print-head of the dot-matrix printer can be addressed individually, lines and curves made up of single dots can be drawn. Figure 4.3(a) is an example of a contour plot produced on a dot-matrix printer. The individual contours are drawn as sets of connected straight-line segments between successive points {$x1,y1$} and {$x2,y2$} that define the digitized contour line. A second example, of an outline map showing the international boundaries of the continent of Africa, is shown in Figure 4.3(b). A pen-plotter normally has built-in firmware that will automatically join two points with a straight line; the command is usually JOIN(X1,Y1,X2,Y2). The user of a dot-matrix printer has to simulate this command by (i) finding the equation of the line joining the two given points {$x1,y1$} and {$x2,y2$}, (ii) using this equation to compute the y coordinate corresponding to each x coordinate in the range $x$min to $x$max, where $x$min is the smaller of $x1$ and $x2$ and $x$max is the larger of $x1$ and $x2$. The equation of a straight line is given by

$$y = a + bx$$

Figure 4.3 *(opposite)* **(a)** Contour map produced using dot-graphics functions on a Mannesman-Tally MT86 dot-matrix printer. **(b)** Outline map of African countries using dot-graphics functions on a Mannesman-Tally MT86 dot-matrix printer.

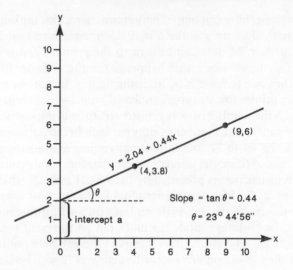

Figure 4.4   Graph of the line $y = 2.04 + 0.44x$. The first term (2.04) is the point at which the line cuts the vertical axis. This value is known as the intercept. The second term is the gradient of the line, given by the tangent of the angle between the x-axis and the line in an anticlockwise direction. The two points (9,6) and (4,3.8) lie on the line.

where x and y are the coordinates of any point on the line, a is the y coordinate of the point where the line cuts the y axis ($x = 0$) and b is the tangent of the angle between the positive x axis and the line, measuring anticlockwise (Figure 4.4). The terms a and b are called the intercept and slope of the line. The slope is found from

$$b = \frac{y2 - y1}{x2 - x1}$$

and the intercept from

$$a = y1 - bx1 \quad (\text{or} \quad a = y2 - bx2).$$

Once a and b are known then the value of y corresponding to every integer value x lying between $x$min and $x$max is calculated from the equation of the line and a dot is printed at the point $\{x,y\}$. The result is a draft contour map which can be produced quickly and cheaply using a simple BASIC program on a small microcomputer equipped with a dot-addressable matrix printer. More efficient ways of drawing lines and circles on dot-addressable (raster) devices such as dot-matrix printers and raster graphics screens are considered by Kingslake (1986).

Coloured maps can be produced on a dot-matrix printer by using a printer ribbon that is divided into three parts horizontally along the length of the ribbon.

The three parts carry either cyan, magenta and yellow ink or red, green and blue ink. The print head can move vertically up or down so that a pin can contact any one of the three parts of the printer ribbon. This type of colour dot-matrix printer is based upon proven and reliable technology; however, the ink on the ribbon can fade if a large map is printed and the result is rather displeasing. A better colour product can be produced using an inkjet printer.

Inkjet printers can be used in exactly the same way as dot-matrix printers. A dot-matrix printer generates a dot on the printed page by striking an inked ribbon with a pin (the pin radius being 1/216 inch) whereas an inkjet printer produces a dot by firing a droplet of ink onto a specially coated paper, the coating being required in order to prevent the ink from running. As the print head moves across the paper individual ink droplets are fired at precise intervals. Characters can be built up as combinations of dots exactly as described above, while contour maps can be drawn by printing individual dots in the same way as a dot-matrix printer in bit-graphics mode. Some inkjet printers use black ink only but colour inkjet printers are becoming a popular choice for cheap colour hard-copy output from microcomputers, as prices start at around £600. The colour inkjet printer uses cyan, magenta and yellow inks (the so-called subtractive primary colours) plus black ink. Red, green and blue (the additive primaries) can be formed by combining cyan, magenta and yellow dots; for example, a magenta dot on a yellow dot gives red because magenta is "minus green" and yellow is "minus blue", so the result is white minus green minus blue, that is, red (Figure 4.5). A black dot can be produced by printing cyan, yellow and magenta dots on top of each other but, since the result is perceived as a kind of muddy brown, it is better to print a single black dot from the black ink reservoir. A white dot is produced by printing nothing at all. A colour inkjet printer can therefore print eight colours (white, black, red, green, blue, cyan, magenta and yellow). While this is adequate for many purposes, more colours can be generated at the expense of a reduction in the number of unit map cells that can be printed across the width of the paper by the use of a procedure called *dithering*.

If a 2×2 pixel matrix of dots is used to represent a single cell on the map to be printed then up to 125 different colours can be printed. Each of the three primary colours is printed separately for each 2×2 matrix, and five levels of each primary colour can be represented (no dots, one dot, . . ., all four dots). Five to the power three is 125, which is the number of colours that can be generated using this procedure. An example of a set of 2×2 matrices showing levels of no dot, one, two, three and four dots is shown in Figure 4.6. For colour output the dot patterns can be rotated as shown to reduce the number of red + green + blue overprints (which are printed as something approaching black). Choropleth maps using different colours to represent different levels of the mapped variable can be generated by specifying one of the 125 colours to be used as the area fill colour. The disadvantage is that the number of horizontal cells is reduced from 760, the number of ink dots that can be printed

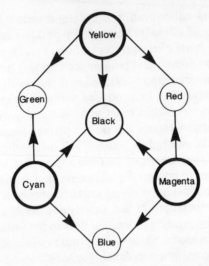

Figure 4.5 The subtractive primary colours of yellow, cyan and magenta can be mixed to produce the (additive) primaries red, green and blue. Each subtractive primary can be thought of as taking away a component from white; thus yellow is "minus blue" and magenta is "minus green" so a mixture of yellow and magenta inks will produce white minus blue minus green – in other words, red. Although an equal mixture of the three subtractive primaries will, in theory, produce black it is normal practice to use a separate black ink.

across one line, to 380 as two ink dots in the horizontal and vertical directions make one cell.

Useful maps can be produced using standard printer technology. The line-printer, dot-matrix printer and inkjet printer each has its own advantages and disadvantages, but all can be used to generate maps ranging from draft-quality

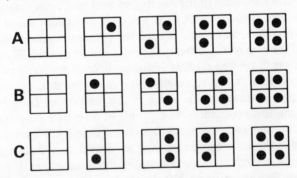

Figure 4.6 Dither matrices for red (A), green (B) and blue (C) components of a colour palette for an inkjet printer. Note that red, green and blue are themselves mixtures of the subtractive primary colours (Figure 4.5). If all three colours overlap (such as in the fifth matrix) then a black dot is printed from the black ink reservoir.

maps such as Figures 4.3(a) and (b) to finished choropleth maps (Figure 4.9). Further details of printer technology can be found in *Byte Magazine* for September 1987.

Dot-matrix or inkjet printers can be used to generate screen dumps or hard copies of the graphics displayed on a colour graphics monitor. A standard dot-matrix printer can show only two states, black and white, but shades of grey can be used to represent colours if 2×2 or 3×3 dot matrices are used to display each single screen pixel. A colour inkjet printer can display eight colours at one dot per screen pixel, though the dithering technique described earlier can be employed to extend the range to 125 colours, though at a lower resolution. An example of a screen dump of a graphics display was used in Chapter 3 (Figure 3.6). The example given here demonstrates the potential use of a simple dot-matrix printer, the Epson RX-80, to produce maps illustrating the use of different map projections.

Map projections are needed in order to display the three-dimensional Earth on a two-dimensional piece of paper (or monitor screen). The transformation from three to two dimensions cannot be achieved without sacrificing one or more of the following properties: area, shape, scale and bearing. Some projections are described as equal-area because the areas measured from maps drawn using those projections are proportional to the true ground areas they represent. To maintain equality of area, however, shape is distorted. Shape cannot be truly represented on a flat map except over small areas but some map projections, described as orthomorphic, have the properties that (i) lines of latitude and longitude intersect at right angles, and (ii) the scale is the same in all directions at a given point, but the scale may differ from one point on the map to another.

Like shape, scale cannot be correct over the whole map, but some map projections can be constructed so that either the lines of latitude or the lines of longitude, or certain lines of latitude and longitude, have the scale correct. Finally, preservation of correct bearing may be important, for example if the map is to be used for navigation. Azimuthal or Zenithal projections can be thought of in terms of the projection of the lines of latitude and longitude on the globe onto a flat sheet of paper which touches the globe at some point (for example, the north or south pole in the case of Polar Azimuthal projections). Bearings from the point of contact (one of the poles) between the sheet of paper and the globe will be correct.

The Mercator projection and the Zenithal Equal-area projection are shown in the form of screen dumps in Figure 4.7. The area shown is Greenland, Baffin Island and Canada/Alaska. Appendix E contains a listing of the data used. The Mercator projection is one of the most common map projections in general use. It was developed in 1569 by Gerhardus Mercator. Lines of latitude are shown on the Mercator projection as being equal in length to the equator so that the scale along these lines of latitude increases away from the equator; so much

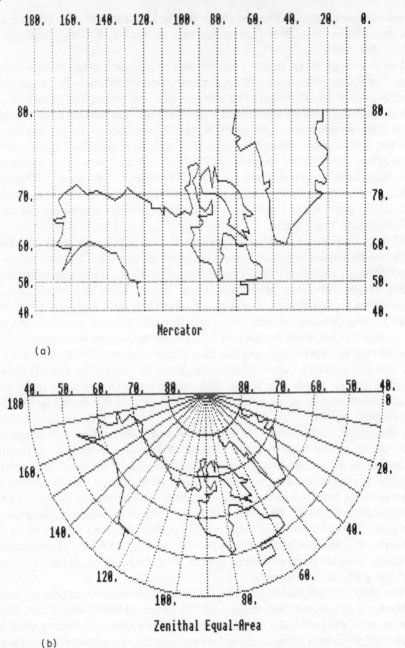

(a)

Mercator

(b)

Zenithal Equal-Area

Figure 4.7   (a) Mercator projection map of Canada, Alaska and Greenland produced using screen dump program on a BBC micro and an Epson RX-80 printer. (b) Zenithal Equal-area map of the same area as (a).

so, in fact, that Greenland appears as large as Canada and Alaska together. The poles cannot be shown because they lie at infinity. However, lines of constant bearing plot as straight lines on a Mercator projection and this is one reason for its popularity; the course of a ship, for instance, could be plotted as a series of straight lines on a map drawn on Mercator's projection.

Many people have a mental map of the world which corresponds approximately to the Mercator projection. They are thus likely to be surprised to find themselves crossing Greenland and Baffin Island if they take a flight from London to Chicago, or passing over Samarkand en route from London to Delhi. Lines of constant bearing (straight lines on a Mercator projection) do not necessarily show the shortest (great-circle) route. A great-circle route is an arc of a circle centred upon the centre of the Earth and passing through the start and end point of the journey; the great circle itself can be thought of as a circular section through the Earth and passing through the Earth's centre. Great circle routes plot as straight lines on some map projections such as the Gnomonic.

In order to draw a map on Mercator's projection the features forming the map must be digitized; that is, the latitude and longitude of a set of points located along coastlines, rivers and other features must be known. The latitude and longitude for a number of points along the coasts of Greenland, Baffin Island and Canada/Alaska are listed in Appendix E. These latitudes and longitudes must be converted into (x,y) plotting coordinates using the formulae:

$$x = R \times \text{longitude}$$
$$y = R \times \log_e(\tan(\pi/4 - \text{latitude}/2))$$

R is the length of the radius of the Earth on the scale of the map and latitude and longitude are given in radians. $360°$ equals 2π radians, and so $\pi/4$ radians in the equation for y is equal to $45°$. BASIC and some versions of FORTRAN contain built-in functions to convert from degrees to radians. Figure 4.7(a) shows the data of Appendix E after (i) plotting on a monitor screen using the formulae given above and (ii) dumping the screen to a dot-matrix printer. Since the scale on Mercator's projection becomes infinite at the north pole, the region beyond 80°N is not shown. Some computers have a PrtSc key; if you press this key the screen display is copied to the printer. However, a lot depends on the hardware available – you must have a printer, at least, and IBM PCs and compatibles require a special graphics card before drawings can be produced on the screen. Special software may also be required for the PrtSc key to work. Figure 4.7 was produced from a screen display on an Acorn BBC microcomputer. The screen display on this machine is "memory-mapped", that is, the contents of the screen are kept in the random-access memory of the computer in binary form (Chapter 1). An assembler-language program reads the screen map and converts each picture element (pixel) to a dot pattern, suitable for printing on an Epson RX-80 printer.

The second example of a map projection shows the same region (Greenland, Baffin Island and Canada/Alaska) as was used for the Mercator example; in the present example the polar case of the Zenithal Equal-area projection is employed. The lines of longitude are straight lines converging on the north pole while the lines of latitude (on the full map) are shown as circles. These circles are spaced so that the area between any adjacent pair of circles (representing lines of latitude) is proportional to the corresponding area on the Earth. The formulae to produce x,y plotting coordinates from latitude and longitude values for this projection are as follows (definitions of r and clong are given first):

$$r = \sqrt{(2R - (R - R \sin(\text{latitude})))}$$

if longitude $> \pi/2$ then

$$\text{clong} = \pi/2 - (\pi/2 - \text{longitude}): m = -1$$

else clong $= \pi/2 - \text{longitude}: m = +1$

$$x = m\ r\ \sin(\text{clong})$$

$$y = -r\ \cos(\text{clong})$$

with latitude and longitude again expressed in radian measure. The screen dump resulting from the application of these formulae is shown in Figure 4.7(b).

Miller and Reddy (1987) give a number of subroutines in the Pascal language for the production of maps using different map projections, and illustrate their use. They also describe three digital data sets. The first is called WORLD.DAT; it contains 6000 points delineating the world's coastlines. These points were derived from (i) the US Geological Survey's "World Outline Map" at a 1:40 million scale, and (ii) the US Defense Mapping Agency's "The World". The second and third datasets are at medium and high resolutions; the high-resolution datasets include international boundaries and consist of 95 000 points and the medium-resolution dataset comprises about 15 000 points. Details of the availability of the datasets and the Pascal programs needed to make use of them are given in Miller and Reddy (1987). Figure 4.8 shows examples of output from Miller and Reddy's programs, the output being produced on an Epson RX-80 nine-pin dot-matrix printer. Snyder (1982) is a useful source of information on map projections, with many worked examples showing the mathematical principles underlying the major map projections.

4.2.2.3 Pen plotters

A pen plotter is a device capable of moving a pen across a sheet of paper with high positional accuracy under the control of a computer. The pen can be either in the down position (when a line will be drawn) or in the up position. The size of the paper is dependent on the type of plotter used; a *drum plotter* has a continuous paper feed (like a tractor-feed printer) with the pen position in the x-direction being controlled by the movement of the entire pen assembly at right-angles to the direction of paper movement. The pen position in the y

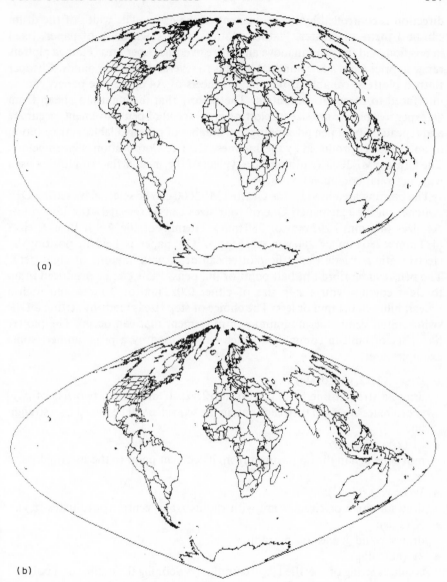

(a)

(b)

Figure 4.8   Examples of screen dumps of maps produced by Miller and Reddy's (1987) program for map projections. **(a)** Sinusoidal projection, an equal-area projection developed in the 16th century. Distortion near the equator and the central meridian is small, but increases considerably towards the poles and the outer meridians. **(b)** Hammer projection, developed in the late 19th century by a German professor of surveying. It is also an equal-area projection and looks more realistic than the Sinusoidal. The screen dumps were produced on an Epson LX-80 printer.

direction is controlled by the movement of the drum. The width of the drum can be 1 metre or greater. Flatbed plotters use a single sheet of paper, fixed in position, and the pen can move anywhere over the paper area. Flatbed plotters range in price from a few hundred pounds to several thousand pounds. Cheaper flatbed plotters, capable of plotting on sheets of A4 or A3 size paper, can be interfaced to personal computers. The factors that distinguish a cheap from an expensive plotter are size of plotting area, speed of pen movement, accuracy and repeatability of pen positioning and number of pens available. A description of an xy plotter produced by the Japanese Graftec Corporation is given below. The Graftec MP3000 xy plotters are typical of the smaller flatbed plotters used with personal computers.

The plotting area used by the Graftec MP3000 can be selected by setting DIP switches on the rear panel. One of four sizes can be selected – the ISO A3 or A4 sizes (420 mm × 297 mm or 280 mm × 216 mm) or the ANSI B or A sizes (432 mm × 280 mm or 280 mm × 216 mm). The paper is held in position by electrostatic attraction and the plotter can draw with any one of eight pens. The pens can be fibre-tip, ball-point or ink pens. Plots can be produced in up to eight colours with a step size of either 0.025 mm or 0.1 mm and with a repeatability of 0.1 mm or less. The choice of step size is made by setting a DIP switch. Step size is the minimum pen movement that can occur. The plotter has a set of built-in commands which can be used by a programmer; some examples are:

- D $(x1,y1,x2,y2)$ (draw):
  draws a straight line from $\{x1,y1\}$ to $\{x2,y2\}$. If more than two sets of $\{x,y\}$ coordinates are given then each pair in the sequence will be joined by a straight line.
- M (x,y) (move):
  move the pen in the up position from its current point to the specified point $\{x,y\}$.
- W (x,y,r) (circle):
  draws a circle or circular arc with radius r and centre coordinates $\{x,y\}$.
- X (axis):
  draws x and y axes on a graph.
- % (hatching):
  connects string of specified $\{x,y\}$ coordinates defining the outline of a polygon and hatches inside it using specified line spacing and angle.
- J (new pen):
  selects specified pen.
- P (print):
  draws specified characters starting at current $\{x,y\}$ position.

These commands can be used to draw maps and graphs by a programmer writing in a high-level language such as FORTRAN or BASIC; the D and M commands are used to draw line segments, given the $\{x,y\}$ coordinates of the start and end of each segment, while the P command is used to add annotation. The % command is useful if the output map is a choropleth map. The GIMMS mapping package, discussed in Section 4.3.2, uses commands similar to those listed above to produce high-quality choropleth maps. Some examples are shown in Figure 4.11.

For some purposes, direct output to photographic film is preferable to paper output. If the map is to be reproduced, for example, a higher-quality end-product is obtained if the map is output onto film for subsequent printing. The alternative is to produce a paper map and photograph it. A device called a *film writer* is available for direct output. The commands to draw lines and characters are much the same as those listed above. In most mainframe mapping packages it is possible to produce the map in what is called *device-independent* form, so that it can be output to any suitable peripheral such as a plotter, a graphics display terminal or a film writer. If this facility is available, the map should be displayed on a graphics display terminal and checked for errors before it is sent to the film writer. The concept of device-independent code is discussed further in Section 4.3.2.

4.3 SOFTWARE FOR COMPUTER MAPPING

4.3.1 Line-printer mapping systems

Some computer centres do not have a plotter mapping package such as GIMMS (Section 4.3.2) available for use. In other cases the plotter hardware is physically located some distance from the user; for example, the user may be in a different building or even at a subsidiary office in another town. He or she may be able to communicate with the central computer via a network (Chapter 1) but special products like graph-plotter output have to be sent through the postal system, and are thus not available quickly. While this delay may not cause difficulties for the production of the final version of a map it is an inefficient way to proceed during the stage of map design. In yet other cases a high-quality plotter-drawn product may not be required, either because the nature of the task is such that a draft product is sufficient or because the cost of producing the required number of plotter-drawn maps would be prohibitive. An undergraduate class of 70–100 students could, for example, so dominate the use of a plotter that other staff and students might effectively be excluded from the use of the device. Hence, even though line-printer mapping systems might at first sight appear to be of little value in comparison with plotter-based systems, there are instances when it is convenient or cost-effective to make use of them.

The SYMAP system is used in this section as an example of a line-printer mapping system; contour and choropleth maps are generated on a line-printer or dot-matrix printer (Section 4.2.2.2). Its development was begun in 1963 by

H. T. Fisher, working at Northwestern Technological Institute, Evanston, Illinois. Later development took place at the Laboratory of Computer Graphics, Harvard University. Its first UK implementation was probably at the University of Edinburgh, Scotland; an early version was brought there in 1967 by Professor J. T. Coppock. One of the most widely-appreciated uses of SYMAP in the UK was Rosing and Wood (1971) who produced an atlas of Birmingham and the Black Country. This atlas was one of the first publications to show the value of computer mapping in geography. SYMAP was made available to the academic community at a nominal charge, thus allowing many research staff and undergraduates the opportunity to become familiar with the concept of computer mapping.

In this section an example of the use of SYMAP to design and produce a choropleth map of the distribution of two variables over 42 countries of the continent of Africa is given. The two variables are infant mortality per 1000 live births and gross national product per head. Both variables are listed in Appendix A and the 42 countries are named in Appendix F. The same variables are used in the statistical examples in Chapter 3 and in the example of the use of the GIMMS plotter mapping system (Section 4.3.2). The examples were produced using the ICL 3900 computer in the Cripps Computing Centre, University of Nottingham. Readers wishing to run the SYMAP program on their local computer system will need to supply different Job Control instructions unless their computer centre is equipped with an ICL computer running under the VME operating system.

SYMAP was developed at a time when interactive computing was still at an experimental stage. Programs and data were, at that time, punched onto cards using a keypunch. Decks of cards, held together (sometimes precariously) by elastic bands were placed in input trays, collected by operators, and fed into the computer through a card-reader. The card deck, together with the output generated by the job, was returned at a later time. Hence SYMAP data formats are 'fixed' by reference to the position of the data element in terms of the 80 columns of the punched card. Since positioning of a character on a video terminal screen is rather more difficult than punching a hole in a specified column of a card, and because today's computer users are accustomed to 'free-format' input (in which data elements are not defined by their position on an input record but are separated by commas or spaces), the input format demanded by SYMAP may seem inflexible and difficult. In the present example the data were entered into a file from a terminal using free format, and a program was written to re-format the data according to the requirements of the SYMAP program. The data used in this example are listed in Appendix G, and are available on an MS-DOS format floppy disc suitable for IBM PCs.

Stage one of a SYMAP run involves the definition of the location coordinates of the spatial units for which the data are measured. In this example the spatial units are the 42 largest countries of Africa (Appendix F). The countries may be referred to as zones or polygons. SYMAP's coordinate system is based on the row/column reference system normally used with matrices. The first coordinate value of a pair refers to position down the page from the top, while the second coordinate is the distance across the paper from the left. The SYMAP coordinate origin is therefore the top left-hand corner of the map area. Most published maps use a grid system and the Cartesian reference system in which coordinates are specified as $\{x,y\}$ pairs. Care must be taken to ensure that the coordinates defining the line segments which are the zone boundaries are specified in row/column terms. If the locational data are available in $\{x,y\}$ order then, if you are familiar with BASIC or FORTRAN, or some other high-level language, it is relatively easy to write a program to reverse the order of the $\{x,y\}$ pairs and to convert the y coordinate so that it measures distance down from the top of the page rather than distance up from the bottom.

Once the zone outline coordinates have been measured and entered into a data file the statistical variable to be mapped is specified. Secondly, the measurements on the statistical variable to be mapped must be presented in the same order that the zone outlines are listed in the data file. Thirdly, details of the map format are given. These three operations – provision of zone outline coordinates, definition of data values for mapping, and specification of map characteristics – are handled by SYMAP through the use of 'packages'. Each package performs a specific function, and the data for each package are provided in a specified, fixed format. The three packages used in this example are called A-CONFORMOLINES, E-VALUES and F-MAP.

A-CONFORMOLINES is used to supply the location coordinates of a set of points defining the position of the boundary of each zone. Unlike the GIMMS system described in Section 4.3.2, the zones are digitized individually, which implies that common boundaries are digitized twice. This can lead to problems if the same coordinates are not entered each time. The zone boundary is considered to consist of a set of straight-line segments the end points of which are required in the A-CONFORMOLINES package. The first and last points are the same so as to close the zone boundary. The use of too few coordinate points will result in angular boundaries while the use of too many points is uneconomical as a printer is only capable of resolving position to within one character position. Countries which have a very small area are not capable of being shown on a large-scale map, so the 42 zones for Africa do not include Djibouti, Swaziland, Lesotho and Mauritius. The coordinate points defining the zones are entered into the SYMAP data file as shown below (refer also to the commands and data for this example, listed in Appendix E). The data must be entered in the columns indicated by the horizontal 'ruler', and whole numbers must be given with an explicit decimal point (i.e. a whole number such as 12 should be entered as 12.0 or simply 12. but not as 12).

```
                     column number
          1111111111222222222233333333334444444445
          1234567890123456789012345678901234567890
          A-CONFORMOLINES
              1     A<vertical><horizont>
                     <vertical><horizont>

                      ....     .....
                      ....     .....
                      ....     .....
                      ....     .....
              2     A<vertical><horizont>
                     <vertical><horizont>

                      ....     .....
                      ....     .....
                      ....     .....
                      ....     .....
          99999
```

The package is introduced by the A-CONFORMOLINES statement. The data for each zone begin with the zone identifying number right-justified in columns 1–5 of the first data record for that zone. Each data record for the zone contains the vertical and horizontal coordinate for a single point; the vertical coordinate is entered in columns 11–20 and the horizontal coordinate in columns 21–30. The last point is the same as the first, so as to close the outline. The points must be entered with the highest point first. The highest point is the one closest to the top of the map. Where two points are equally high the leftmost one should be chosen. Once the first point has been identified the remaining points are entered in clockwise order around the zone boundary. The A-CONFORMOLINES package is terminated by the code 99999 in columns 1–5.

E-VALUES are the data values for the zones to be mapped. These values are entered in the same order as the zone boundary definitions in the A-CONFORMOLINES package. The World Data Matrix (Appendix A) contains data for the largest 100 countries of the world and some of the African countries whose borders have been digitized are not included. The value −1 (minus one) is used to indicate that no data are available for certain countries (specifically: Mauritania, Guinea-Bissau, Liberia, Gabon, Congo, the Central African Republic, Namibia, Togo and Botswana). The E-VALUES package is introduced by the characters E-VALUES in columns 1–8 of a data record. The data values for each zone are then given in order in columns 11–20, with one data value per record. An example is shown in Appendix G. The data are followed by the code 99999 which signals the end of the package.

The F-MAP package allows the user to choose from a number of options, called "electives" in SYMAP language. There are too many to list here; the

SYMAP documentation, available from your computer centre, provides a full account. The options used in this example are:

- Elective 1: specification of vertical and horizontal physical dimensions of the printed map, in inches (1 inch = 2.54 cm).
- Elective 2: size of the printed map in terms of the map coordinate system.
- Elective 3: number of classes into which the data range is to be divided.
- Elective 4: minimum value in the data. This elective can be used to ensure that those data with a value below that specified by elective 4 are mapped. Zones (countries) with values less than the minimum are allocated the symbol 'L' on the printed map.
- Elective 5: as elective 4 except that the maximum required value is specified. Countries with data values greater than this maximum value are allocated the symbol 'H' on the printed map.
- Elective 6: class interval specification. If elective 6 is chosen without any parameters then the range of the data (or the restricted range given by one or both electives 4 and 5) is divided into classes such that there are an equal number of countries in each class. If this elective is not requested then the data range is divided into classes with an equal range. The number of classes is five unless the user opts to change this value.
- Elective 8: normally SYMAP leaves a gap around the boundaries of each country. If elective 8 is specified then this gap is omitted.
- Elective 23: use of this elective will cause SYMAP not to print an invalid data point symbol for countries without data.

Figures 4.9(a)–(c) show the output from a SYMAP run using the commands and data of Appendix G. Figure 4.9(a) uses the default five classes in a map of infant mortality per thousand live births for the African countries included in the World Data Matrix (Appendix A) and named in Appendix F. Data values of − 1 are excluded by the use of elective 4. The value − 1 is used to flag missing data; the countries concerned are shown on the map by the symbol 'L'. The five classes are printed using the following symbolism class 1 (.), class 2 (+), class 3 (O), class 4 (O and − superimposed) and class 5 (O, X and M superimposed. Figure 4.9(b) uses the same data as Figure 4.9(a) but elective 8 has been used to eliminate the blank lines around the country boundaries. The third map shows the distribution of values of Gross National Product in American dollars for the same countries. Elective 4 has again been used to map countries for which no data are available. Elective 5 has also been selected; this elective limits the upper range of the data. The GNP per head of Libya is far greater than that of any other African country, so elective 5 was used to cause SYMAP to allocate the symbol 'H' to Libya on the printed map. The data between the lower and upper boundaries specified by electives 4 and 5 are mapped so that each class has, as far as possible, an equal number of countries. These maps should be compared with Figures 4.11(a)–(c), which are the output from the plotter mapping package GIMMS, which is described in Section 4.3.2.

Once the A-CONFORMOLINES and E-VALUES data have been prepared it is a simple matter to use different electives in the F-MAP package in order to alter particular settings, such as the number of shading levels, the choice of class interval (equal width, equal number of zones or user-specified), or the type of symbolism used to define the shading pattern. The cost of running SYMAP is low since standard peripherals are used, so it makes sense to use it as an aid to map design before proceeding to a more elaborate plotter mapping package such as GIMMS. The major disadvantage of doing so is that the data formats required by the two packages are very different. In particular the way in which the zone boundaries are digitized is not compatible. However, a simple data

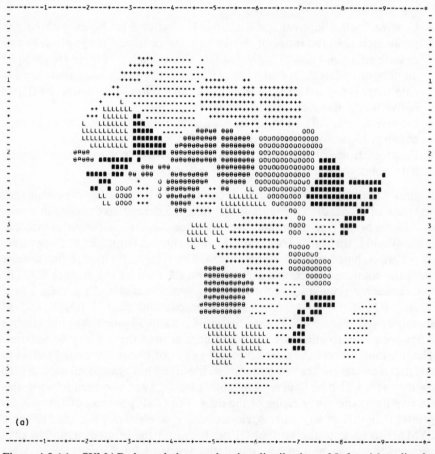

Figure 4.9 **(a)** SYMAP choropleth map showing distribution of Infant Mortality for 42 African countries (Appendix F). Blank spaces separate the countries. Increasing density of shading corresponds to greater rate of infant mortality. Symbol L indicates no data.

reformatting program may be available at your computing centre to help you convert data from one format to another. My own data format convertor takes as input a set of digitized points and allows the user to specify the order in which the {x,y} coordinates are written to a disc file. Operations such as "reverse the order of x and y" and "subtract the y coordinate from ymax" can be given in order to allow the easy conversion of locational (coordinate) data from GIMMS format (Cartesian coordinate format) to SYMAP (row/column format). The disc file produced by this format convertor program is then edited using a standard text editor.

4.3.2 Plotter packages

GIMMS, developed by Mr. T. C. Waugh of Edinburgh University, Scotland, is used in this section as an example of a widely-used plotter mapping package. As in the case of the SYMAP examples earlier in this chapter, the facilities

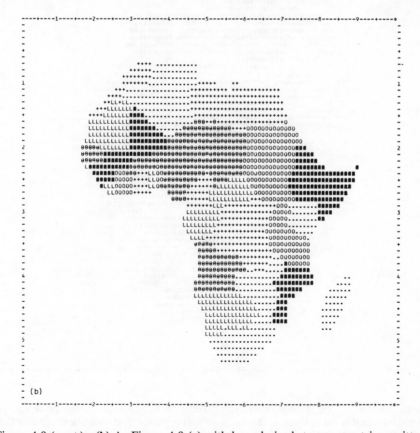

Figure 4.9 (*cont.*)   (b) As Figure 4.9 (a) with boundaries between countries omitted.

of the Cripps Computer Centre at Nottingham University were used and specific examples of Job Control commands will require modification to suit the requirements of your computer system.

The first task in the process of producing a computer-drawn choropleth map using GIMMS is to input the digitized map (Appendix H) as a set of line segments joining nodes (as described in Chapter 2) and to link these segments together to form polygons. The data in Appendix H were digitized manually using an *xy* grid with coordinate values in the range 0–1000. The coordinates of each point were recorded to the nearest 5 units. Each line segment was defined by

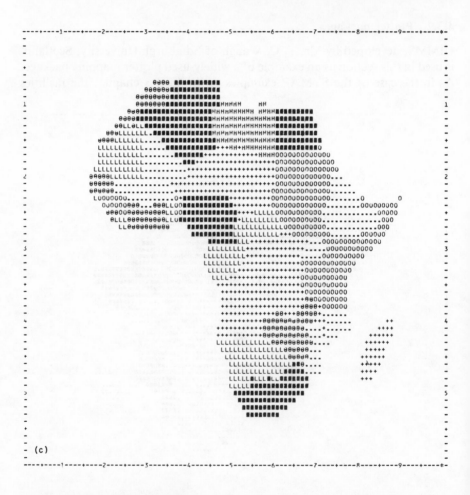

Figure 4.9 (*cont.*)   (c) Distribution of Gross National Product for 42 African countries. The symbolism is interpreted in the same way as Figure 4.9(a) with the exception that the character H indicates a value beyond the upper limit specified by elective 5.

a sequence of points joining two nodes, the nodes being points where two or more segments join. Next, the *xy* coordinates of the points making up each line segment were read from the map and recorded manually, together with the identifier of the country to the left and the country to the right in the direction of digitizing (as illustrated in the 2D encoding and USGS DLG examples of data structures in Chapter 2). The identifier SEA was used if there was no country to the right or left of the line segment. The full description of a line segment for input to GIMMS is:

- identifier of country to the left of the line
- identifier of country to the right of the line
- *xy* coordinates of points defining the line
- terminator

The terminator is the backslash character / (ASCII code 47). The "country left" and "country right" items are the country identification numbers from Appendix F preceded by the letter Z (meaning zone).

The following GIMMS commands were entered into a file called (on the Nottingham University ICL 3900 computer system) PMMLIB.GIMMSAF. They perform the following procedures:

(i)   input the digitized map data to GIMMS, and
(ii) generate the linkages between line segments and so identify which line segments bound which countries:

```
*FILEIN SEGMENT FILEOUT = 10 FILENAME = AFRICA
TITLE = 'AFRICAN COUNTRIES GIMMS SEGMENTS'
LIMITS 0,0,1000,1000
BEGIN
SEGMENTS
⟨. . . GIMMS line segment descriptors as described
   above and listed in Appendix H . . .⟩
END
*POLYGON    FILEIN = 10 FILEOUT = 11 ALPHA EXCLUDE ZONE = SEA
*STOP
```

The *FILEIN command tells GIMMS that a data file is being input. The qualifier SEGMENT indicates that this is a file of line segments. FILEOUT = 10 means that the line segment file will be stored in a file connnected to unit 10 (unit connections are specified in the Job Control commands, which are described below). TITLE can be used to add a descriptive comment to help identify the file at a later stage. LIMITS gives the *xy* coordinates of the lower left and upper right corners of the map area. The line segment data are introduced by the words BEGIN and SEGMENTS and are terminated by the word END. At this stage, when the commands are executed, the line segment data will be stored in a file connected to unit 10. The *POLYGON command is a GIMMS instruction to fetch the file whose unit number is 10, read from it the line segment data, and link the line segments together to form polygons. The polygon definitions are written to the file whose unit number is 11. ALPHA tells GIMMS to store the zones in alphabetic order of their identifiers, so that Z01 comes before Z02, and so on. This is important because the statistical data to be mapped must be associated with the correct zone. EXCLUDE ZONE = SEA is used to eliminate from consideration the zone whose identifier is SEA. This identifier was used whenever the zone to the left or to the right of the line segment being digitized was sea. SEA can be considered as the identifier of the zone that surrounds the digitized map of Africa.

The GIMMS commands and line segment data described above are stored in a file called PMMLIB.GIMMSAF on the Nottingham University ICL 3900 computer system, as described in the preceding paragraph. This computer system, like other multi-user mainframes, requires that each job or task be preceded by introductory commands and followed by terminating commands. In the case of the Nottingham system the Job Control commands needed to define the GIMMS task are:

JOB(NULG.LGPMMGIMMS1, STA = 3)
ASSIGN_OUTPUT_FILE(PMMLIB.GIMMSPOLY,OPEN_ON = 10)
ASSIGN_OUTPUT_FILE(PMMLIB.GIMMSPOLY1, OPEN_ON = 11)
GIMMS(INPUT = PMMLIB.GIMMSAF)
ENDJOB

The JOB command tells the operating system of the ICL 3900 which individual is running this job (PMM), what his university (NU), faculty (L) and department (G) are, the name of the job (LGPMMGIMMS1) and the priority to be given to the job (3 on a scale 1–5). The JOB command will differ from one computer to another, as will the next two commands which relate filenames to unit numbers. The ICL VME operating system command to relate the file PMMLIB.GIMMSPOLY to unit number 10 is ASSIGN_OUTPUT_FILE. Two output files are created by the two ASSIGN_OUTPUT_FILE commands. One (on unit 10) will hold the line segment data and the second, opened on unit 11, will contain the polygon definitions. The polygon definition file will be used in a later program run when the polygons will be associated with statistical values (one of INFMORT or GNP in these examples). The GIMMS command tells the VME operating system to activate the GIMMS program, using the instructions contained in the file PMMLIB.GIMMSAF. The Job Control instructions terminate with the ENDJOB command.

The Job Control commands are themselves placed in a file, using a text editor. I used the fileneame PMMLIB.GIMMSJOB1 to hold the Job Control commands. This file is then sent to batch queue number 3 (because STA = 3 was entered on the JOB command line). Batch queues are described in Chapter 1. The ICL VME operating system command to enter a file of Job Control commands into a batch queue is SUBMIT_JOB or SBJ for short, so the command

SBJ(PMMLIB.GIMMSJOB1)

entered on a terminal will ask the VME system to place the commands held in the file PMMLIB.GIMMSJOB1 into the queue number specified on the JOB command line. Other computer operating systems have a different but synonymous command, for example SUBMIT on a DEC VAX/VMS system. The jobs in the batch queue are executed when they reach the head of their particular queue, and the output is sent to the system line-printer.

A sample of the printer output from this GIMMS job is reproduced as Figure 4.10. Each polygon or zone is identified by its zone identifier (*Znn* in Appendix H) and the number of bounding line segments is given. The ENVELOPE is the minimum rectangle containing the polygon, and the coordinates following the word ENVELOPE are the lower left and upper right corners of this rectangle. Finally, the area and the coordinates of the centre of gravity of the zone are output. Note that the area is not the actual ground area of the zone but is expressed in terms of the units on which the coordinates were measured. At the end of the output the total area of the polygons and the number of points processed is given. In this example the total area is 365 825 square units and the number of points processed was 330.

A second file is printed by the VME operating system for all batch jobs. This file is called the log file. It records the progress of the job, including error messages, and also holds messages from the system manager (details of system availability, announcements of problems, and so on). The log file for my GIMMS job tells me that the job reached the head of the batch queue and started to execute at 48 minutes and 56 seconds after 5 o'clock p.m. on Monday, January 4th, 1988 and ended at 49 minutes and 15 seconds past the same hour on that day.

Once the polygon definition file has been created, a second file, containing the statistical data to be mapped onto the polygons, can be generated. In the present example, this file contains data extracted from the World Data Matrix listed in Appendix A for the countries named in Appendix F. Two variables, infant mortality and Gross National Product, identified as INFMORT and GNP respectively, are used. Since data for some of the smaller countries are not incorporated in the World Data Matrix the value − 1 (minus one) is used to indicate "no data available". The GIMMS commands to build the statistical data file are:

```
POLYGON = Z01                SEGMENTS = 3
ENVELOPE (MIN X,Y : MAX X,Y) : 135 760 335 930
AREA=8600.0   CENTROID=252,855    POINTS=13

POLYGON = Z02                SEGMENTS = 7
ENVELOPE (MIN X,Y : MAX X,Y) : 245 710 465 930
AREA=27312.5  CENTROID=366,824    POINTS=14

POLYGON = Z03                SEGMENTS = 3
ENVELOPE (MIN X,Y : MAX X,Y) : 425 850 470 930
AREA=2075.0   CENTROID=446,893    POINTS=7

POLYGON = Z04                SEGMENTS = 7
ENVELOPE (MIN X,Y : MAX X,Y) : 440 720 610 880
AREA=19300.0  CENTROID=528,806    POINTS=13

POLYGON = Z05                SEGMENTS = 4
ENVELOPE (MIN X,Y : MAX X,Y) : 600 745 725 860
AREA=11775.0  CENTROID=659,799    POINTS=8

POLYGON = Z06                SEGMENTS = 5
ENVELOPE (MIN X,Y : MAX X,Y) : 130 675 280 825
AREA=14150.0  CENTROID=212,741    POINTS=13

POLYGON = Z07                SEGMENTS = 5
ENVELOPE (MIN X,Y : MAX X,Y) : 115 640 185 700
AREA=3175.0   CENTROID=149,669    POINTS=8

POLYGON = Z08                SEGMENTS = 3
ENVELOPE (MIN X,Y : MAX X,Y) : 125 620 160 645
AREA=387.5    CENTROID=142,635    POINTS=4

POLYGON = Z09                SEGMENTS = 7
ENVELOPE (MIN X,Y : MAX X,Y) : 495 60 695 210
AREA=15737.5  CENTROID=601,130    POINTS=15

            .              .
            .              .
            .              .
            .              .
            .              .


ZONES CHECKED = 42


AREAFILE '' WITH 42 ZONES CREATED ON CHANNEL 11
SUM OF AREAS = 365825.00
NO. OF INPUT POINTS = 330


COMMAND:*STOP


---- END OF GIMMS RUN ----
```

Figure 4.10   Extract from printout generated by the GIMMS program.

*FILEIN DATAFILE ZONES = 42 VARS = 2
   NAMES = INFMORT, GNP
   ⟨- - - data for 2 variables on 42 polygons entered in zone order, i.e. the data
   for the polygons are specified in the order Z01, Z02 . . . Z42 where Z01,
   Z02 . . . Z42 are the polygon codes used in the program to input the line
   segments defining the polygons. Where data for a particular zone are not
   available the code − 1 is used. - - -⟩
*SAVE DATA TO FILE 12
*STOP

The Job Control commands given below assume that (i) the instructions
and data for the statistical data definition are held in a file called
PMMLIB.GIMMSTATDAT and (ii) the GIMMS-generated data file will
be written to another file called PMMLIB.GIMMSPOLYDAT which
will be opened on unit 12. The ICL VME Job Control commands are:

JOB(NULG.LGPMMGIMMS2,STA = 3)
ASSIGN_OUTPUT_FILE(PMMLIB.GIMMSPOLYDAT,OPEN_ON = 12)
GIMMS(INPUT = PMMLIB.GIMMSTATDAT)
ENDJOB

This Job Control file is sent to the batch queue by the SUBMIT_JOB
command as described above. The output from GIMMS consists of a list
of the options specified together with informative messages to the effect
that a data file with 42 zones and 2 variables has been created and saved to
the file opened on unit 12.
   On completion of this second GIMMS job we are in a position to begin the
final task, that of producing maps based on the digitized map of the 42 countries
of Africa, using one or other of the two statistical variables entered into the
file PMMLIB.GIMMSPOLYDAT. The GIMMS commands to produce a
choropleth map are listed below. The numbers in square brackets are NOT part
of the GIMMS command. They are identification numbers that are used in the
description below.

*PLOTPARM PLOTTER   [1]
 *PLOTPROG   [2]
  *NEWMAP MAPSIZE = 25,25 FRAME   [3]
  *GIMMSFILE FILE = 11   [4]
  *RESTORE DATA FROM FILE 12   [5]
  *TEXT   POSITION = 12.5, 24   [6]
          SIZE = 0.5   [7]

ALPHABET = 61   [8]
CENTRE ONX   [9]
'Infant mortality mid-1980s'   [10]
*MAP VARIABLE = INFMORT, TYPE = AREA   [11]
  *END   [12]
*STOP   [13]

The GIMMS command file is stored in a file on the ICL 3900 computer under the name PMMLIB.GIMMSMAP1 and the ICL VME Job Control instructions required to place this command file into batch queue number 3 (STA = 3) are:

JOB(NULG.LGPMMGIMS3, STA = 3)
ASSIGN_DATA_FILE(PMMLIB.GIMMSPOLY1, OPEN_ON = 11)
ASSIGN_DATA_FILE(PMMLIB.GIMMSPOLYDAT, OPEN_ON = 12)
GIMMS(INPUT = PMMLIB.GIMMSMAP1,DIG_FILE =
                                    PMMLIB.GIMMSDIGFILE1)

This Job Control file tells the VME operating system that two data files are required, one (created by the first GIMMS program) is called PMMLIB.GIMMSPOLY1 and is to be opened on unit 11. This file holds the polygon (zone) outlines in order Z01 to Z42. The second data file is the one holding the statistical data for the 42 polygons for the variables INFMORT and GNP. Its name is PMMLIB.GIMMSPOLYDAT and it is to be opened on unit 12. The GIMMS statement specifies that the GIMMS commands are to be read from (INPUT =) the file PMMLIB.GIMMSMAP1, which is listed above. The second instruction on this line of the Job Control file (DIG_FILE =) will be discussed later. First, however, a description of the instructions contained in the file PMMLIB.GIMMSMAP1 is given. These instructions are to be executed by the Job Control instructions that are listed immediately above. The following paragraphs are numbered so as to correspond with the records of the file PMMLIB.GIMMSMAP1.

1. *PLOTPARM PLOTTER indicates that the map is to be output to a pen plotter.
2. *PLOTPROG begins the instructions to plot the map. The commands labelled 3 to 12 form the PLOTPROG package.
3. *NEWMAP means that this is a new map. MAPSIZE = 25,25 gives the x and y dimensions of the map in centimetres. If the y-dimension of the map exceeds 29 cm the map will be drawn on wide paper and the computer operator will have to change over the paper on the plotter. Narrow (29 cm) paper is generally used for draft maps to reduce costs. FRAME causes a rectangle to be drawn around the map.
4. *GIMMSFILE FILE = 11 associates the polygon outline file with the file PMMLIB.GIMMSPOLY1 opened on channel 11 in the Job Control file.

5. *RESTORE DATA FROM FILE 12 reads the statistical data from the file PMMLIB.GIMMSPOLYDAT opened on channel 12 in the Job Control file.
6. *TEXT is used to provide annotation on the map. POSITION = 12.5,25 gives the x,y coordinates of the text, in centimetres, with reference to the bottom left-hand corner of the map. In this case, the command CENTRE ONX (number 9 below) is used so the annotation will be centred on a position 12.5 cm along the x-axis.
7. SIZE = 0.5 gives the height of the characters to be used, in centimetres.
8. ALPHABET = 61 specifies the type of lettering. A range of character fonts is available in GIMMS; set number 61 (an open alphabet) is used here. The others are listed in the GIMMS documentation.
9. CENTRE ONX causes the text to be centred on the x coordinate specified by the POSITION command (number 6 above).
10. *MAP VARIABLE = INFMORT, TYPE = AREA. This is the command to generate a choropleth map for the variable INFMORT, using the parameters specified above.

The map generated by these commands is not sent directly to the plotter. Instead, the map is stored in digital form in the file PMMLIB.GIMMSDIGFILE1 which is associated with the sub-command DIG_FILE = in the Job Control command 'GIMMS'. The term DIG_FILE means Device Independent Graphics File, which is a digital map held in a disc file. The term "device-independent" means that the digital map can be drawn on any available graphics output device; at Nottingham a graphics screen or a plotter can be used. The data contained in the DIG file can be sent to a graphics screen by the command "VIEW_DIG_FILE". This is the recommended choice for the first viewing of a DIG file, for the display process is much more rapid than if the data are output directly to a plotter. Also, the map can be regenerated if it is apparent that a mistake has been made in the specifications. When the user is satisfied that the map meets his or her requirements the contents of the DIG file can be sent to the plotter using the "PLOT_DIG_FILE" command. Remember that these system commands are specific to the Nottingham University ICL 3900 computer system; you will have to use the equivalent command appropriate to the computer system used at your computer centre.

Figure 4.11(a) shows the result of the procedure described above. Although the map shows the distribution of the variable INFMORT over the countries of Africa, it is not satisfactory for two reasons. Firstly, it is not clear that data are unavailable for some of the countries. Secondly, the class intervals for the five classes used in the choropleth map have been derived simply by dividing the data range into five equal ranges. A second map (Figure 4.11(b)) was easily generated to overcome these two problems. The areas for which data are not available were excluded by specifying that values less than zero should not be shaded but should be left blank. This is done by using the subcommand MINIMUM = 0 as shown below. The second problem was overcome by specifying TYPE = QUANTILE. There is a range of such subcommands available in GIMMS; these two were selected to illustrate the way in which changes can easily and quickly be made to a command file if the result of a GIMMS run does not meet the user's requirements. Further details of the commands available are given in Carruthers (1985).

The additional command needed to specify that areas with a negative data value be excluded and that the class intervals be chosen so as to place an equal number of zones in each class is:

*INTERVALS VARIABLE = INFMORT MINIMUM = 0 TYPE = QUANTILE

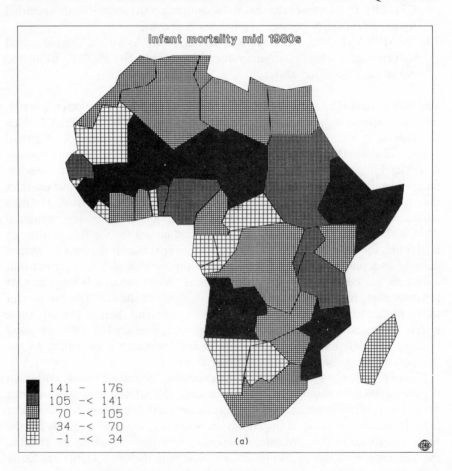

Infant mortality mid 1980s

■	141 – 176
	105 –< 141
	70 –< 105
	34 –< 70
	–1 –< 34

(a)

Figure 4.11 **(a)** GIMMS plotter map of infant mortality for 42 African countries using equal class intervals. No special provision is made for missing data.

This command is placed after the *TEXT command and immediately before
the *MAP command, that is, between lines [10] and [11] in the command
file listed above. The result is a more suitable map (Figure 4.11(b)) on which
it is clear that data are unavailable for some countries. The five shading symbols
are also more equally spread over the map area.

The legend in the lower left corner of the map is called a default option. This
means that the user does not have to specify that a legend is required, nor is
it necessary to give any detailed instructions on the placement of the legend.
However, the default can be overridden either by (i) specifically requesting that
no legend be plotted or (ii) giving a complete specification of the required legend.
Many other GIMMS options have default settings which are used if no alternative
is specified (or if the option is not specifically included in a command file).

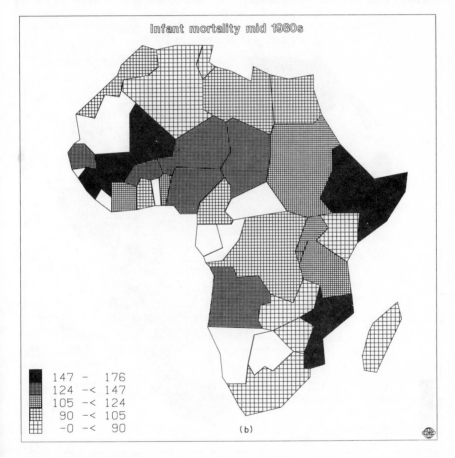

Figure 4.11 (*cont.*) (b) Same data as Figure 4.11(a) but with countries without data
being left blank. The classes contain approximately equal numbers of members.

The map for the second variable, GNP, is shown in Figure 4.11(c). Again, a minimum data value of zero is specified so as to eliminate those countries for which no data are available. The option TYPE = QUANTILE is also used. The resulting map can be visually compared with Figure 4.11(b) in order that any similarities in the spatial distributions of infant mortality rate and Gross National Product can be detected.

These examples illustrate only some of the potential of the GIMMS package for the presentation of geographical information in map form. Other mapping systems may well be available at your local computer centre. Whatever computer mapping system is used, the major advantage of such systems over manual map production quickly becomes apparent, namely, the ease with which maps can be redrafted and revised once the basic polygon data and statistical database have been entered into the computer. The mechanics of map drawing are no

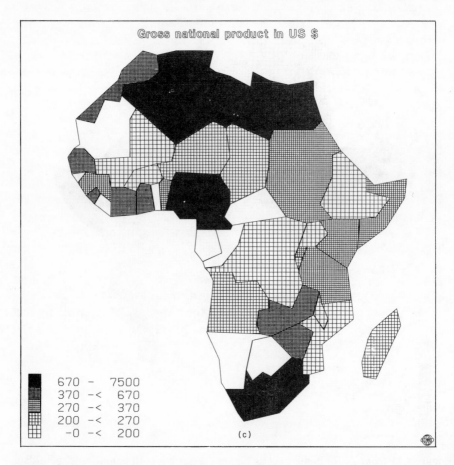

Figure 4.11 (*cont.*)   (c) As Figure 4.11(b) using data for Gross National Product.

longer a constraint or a deterrent to the use of maps for the user can, given a suitable package such as GIMMS, modify and refine his or her map or generate a number of maps without difficulty.

4.4 SUMMARY

Automated cartography, computer-aided cartography and digital mapping are different names for essentially the same process, the generation of maps and diagrams by computer. In this chapter, hardware for automated cartography has been described. The graphics terminal, various types of printer (line-printer, dot-matrix printer and inkjet), plotter (drum, flatbed), and film-writers have been described in terms of their suitability for output of maps and diagrams. Each is suitable for a particular application, though there is some overlap. The cheapest and most readily-available output device is the printer. Dot-matrix printers, such as those used with popular makes of personal computer, and line-printers (which are normally used with larger, usually centralized, computers) can be used to generate cheap, but not particularly accurate, maps and diagrams. They are most useful in education, because students are able to get a fast turn-round while at the same time learning the principles of operation of a particular software package, and in draft map production, where a cartographer is experimenting with different map designs. Inkjet printers provide a similar facility but with the added advantage of colour.

The quality of the display produced by a graphics terminal depends upon the horizontal and vertical resolution of the screen, expressed as the number of phosphor dots in the two directions, and on the number of colours that can be simultaneously displayed. Modern graphics terminals have a screen resolution of 1024×1024 or more, and can display eight or more colours on the screen at the same time. These devices are useful for experimenting with different map designs and for the production of "temporary" maps. An example of the latter use might be where a sequence of maps showing alternative strategies (such as routes for a motorway) was required for demonstration purposes, for example at a public enquiry. A graphics display screen could be used to switch rapidly from one map to another; this would have the advantage of economy (in that costly high-quality paper maps would not be needed by each participant) and would be one way of ensuring that everyone was looking at the right map at any particular time. In-car maps of possible alternative routes are another example of "temporary" maps, which have no permanent value. Maps shown on a graphics screen can be output to a dot-addressable printer using a screen-dump program.

High-quality pen-and-ink maps are produced on a graphplotter. There is a wide range of such plotters available, from the cheapest A3 or A4 paper size plotter designed for use with a personal computer to the highly-accurate large flat-bed plotter. Where a map is required to be printed (for example in a book or a report) it can be output to a film-writer which can produce a film negative. This negative can be used directly in the printing process.

Software is required in order to make the hardware work. Since a wide range of different software packages is available, two of the more widely-used packages have been chosen to demonstrate the way in which such packages are used. SYMAP is the best-known line-printer mapping package; it has a large number of users in universities, colleges and research institutions. SYMAP has the advantage that no specialized hardware is required and is thus useful for introductory teaching at degree level. Its other uses are in experimenting with different map designs. GIMMS is a plotter mapping package capable of producing professional-quality output. The examples of SYMAP and GIMMS in this chapter use the same data – infant mortality rates and Gross National Product for 42 African countries – so the reader can judge the comparative merits of output from the two systems.

There are several reasons why computer mapping is increasing in popularity. First of all, more spatial data are becoming available in digital or computer-readable form. The map user can select features of interest from the digital map and display them in a way that is suited to his or her requirements. Constraints set by different map projections, or by map borders, can be overcome by the computer. Secondly, computer mapping allows the user to experiment; if the final product is not satisfactory then the labour involved in redrafting the map is negligible in comparison with manual methods. Thirdly, digital maps can be combined with other kinds of spatial data in a Geographical Information System (Chapter 7). A computer-drawn map is generally the end-product from such a system, which allows the user to combine map data with other kinds of data about places (economic, demographic, geological and pedological data, for instance) in order to produce cartographic and tabular output to suit a specific purpose.

The use of computers in cartography has had a number of effects; first of all, cartography as a discipline has become more independent of geography. Whilst there is no denying that cartography in its more advanced forms was never seen as an integral part of geography, it is nevertheless true that as cartography adopted more advanced technology it became more remote from geography, possibly as a result of the relatively lower interest in technology among most geographers. Secondly, the use of computers has sparked off an interest in the development of new, more clearly-defined methods in cartography while, at the same time, relieving the cartographer from the labour of manual drafting. More thought and time is given to questions of map design and methods of data manipulation. These, and other, aspects of the effects of the adoption of computer-assisted methods on cartography are described by Morrison (1980). Further recommended reading is Peucker (1972) and the *Transactions of the Institute of British Geographers* Special Issue on Contemporary Cartography (volume 2, number 1, 1977).

4.5   REVIEW QUESTIONS

1. Define the following:

   GPS
   master survey drawing
   screen resolution
   dot graphics
   screen dump
   elective

   analogue map
   interlacing
   choropleth map
   dither matrix
   map projection
   device-independent graphics

   screen refresh
   subtractive primary
   orthomorphism

2. What, in your view, are the advantages and disadvantages of computer mapping compared to manual methods?

3. List the steps you would follow in preparing the input for SYMAP, GIMMS or any other computer mapping package with which you are familiar.

4. What is meant by "hardcopy"? Give examples of hardcopy devices suited to the production of cartographic products, listing the properties of each device.

5. State the properties of the Mercator and Zenithal Equal-area map projections. Which projection would you choose for a map to be used by a yachtsman sailing from Rio de Janeiro to Liverpool?

CHAPTER 5

Remote Sensing

5.1 INTRODUCTION

Remote sensing is the collection of information about the properties of an object without physical contact between the observer and the object being made. It is an everyday experience; our eyes, ears and noses collect information about distant objects. On the other hand, our senses of taste and touch do not act remotely. In environmental remote sensing the properties of interest are those of the Earth's surface and atmosphere, and the data collected by remote sensing programmes are used to infer the nature of the Earth's surface and atmospheric features. Such properties are of interest to a range of practical applications including mapping and surveying, weather forecasting, agricultural crop yield estimation, forest management and geological exploration.

In some of these applications, remotely-sensed data in the form of air photographs are used. These photographs are interpreted visually using a variety of instruments. Over the last three decades digital or numerical images of the Earth's surface and atmosphere have been obtained from imaging instruments carried by satellites. Satellite-borne instruments are capable of observing large areas of the Earth's surface on a repetitive basis, thus providing large area regional or synoptic coverage over time. Such data are valuable for detecting and monitoring change, for example in the type and cover of vegetation, and for observing dynamic phenomena such as sediment patterns in estuaries and coastal waters, or the movements of warm and cold fronts in the lower atmosphere.

Some satellites remain stationary relative to the Earth. The instruments that they carry are thus able to view an entire hemisphere and to return images every 30 minutes. These satellites are in *geostationary* orbit and the images that they provide are used primarily in weather forecasting applications. Figure 5.1 shows an image acquired by the Pretoria (South Africa) receiving station from the European geostationary satellite Meteosat. Other satellites are in a *near-polar* orbit which takes them over the Arctic and Antarctic regions. They circle the Earth repetitively and, as the Earth rotates eastwards below their orbital path,

140

PRODUCED BY THE SATELLITE APPLICATIONS CENTRE OF THE C.S.I.R.
METEOSAT V2   IMAGE   DATE: 02 AUG 1988 AT 12H00

Figure 5.1   Remotely-sensed image showing Africa and Europe. This image was collected by the European geostationary meteorological satellite Meteosat, and acquired by the South African ground receiving station near Pretoria (inset). Images such as these, which are collected every 30 minutes, are routinely used in weather forecasting and in atmospheric modelling.

their instruments build up a picture of land and sea-surface conditions and the state of the atmosphere. The pictures on the evening TV weather forecast come from the US NOAA satellites which orbit at an altitude of around 850 km. From this altitude their instruments view a swath of the Earth's surface that extends 1500 km to either side of their suborbital tracks, so that a picture of the globe over a 24-hour period is built up. The Landsat satellites, which are operated by the American Eosat company, have an orbital altitude of 705 km and their sensors have a narrower field of view which covers the area within 92.5 km of their suborbital tracks. It takes 16 days for Landsat's instruments to build up a global picture. However, Landsat's imaging instruments are capable of showing finer detail of surface features than are the instruments onboard the NOAA satellites, and so Landsat images are used in studies of agricultural crops, coastal sediment patterns and other fields in which high resolution is required. NOAA images are used in weather forecasting which requires frequent global coverage but at lower resolution. Data from Landsat's instruments are used in this chapter to illustrate geographical applications of remote sensing. Other satellite remotely-sensed data, such as the NOAA images already mentioned and the images from the French SPOT and the Japanese MOS satellites, can be processed by the methods described and illustrated in this chapter.

5.2   LANDSAT SATELLITES

The first Landsat satellite was launched in 1972. Two further satellites, Landsats 2 and 3, were launched in 1975 and 1978 respectively. These were the "Mark 1" Landsats, and all three have now been retired from service. The first of the "Mark 2" Landsats, numbered 4, was launched in 1982 and the second (Landsat 5) in 1984. Both Landsats 4 and 5 are still operational in late 1989, though Landsat 4 is operating on reduced power. Landsats 1–3 and Landsats 4 and 5 have rather different characteristics. The earlier series will be described first.

Landsats 1–3 were placed in a near-polar orbit at an altitude of 910 km. The later Landsats, 4 and 5, are in a lower orbit at 705 km. Both orbits are described as *sun-synchronous* because the relationship between the positions of the satellite, the sun and the Earth is maintained as the satellite passes over the illuminated side of the Earth, so that the local sun time on the ground immediately below the satellite is approximately the same, around 9.30 am. Landsat images of a given area will therefore always be taken at the same local solar time and not at varying times throughout the day, thus maintaining a constant direction of illumination. This factor is an important one if images of an area taken on different days are to be compared. Figure 5.2 shows the orbit of Landsats 4 and 5, and it is apparent that the satellites do not pass directly over the north and south poles; the furthest north they reach is 82°N and the furthest south is 82°S. Every point on the Earth's surface between these two latitudes was imaged at least once every 18 days by Landsats 1–3, and is imaged every 16 days by Landsats 4 and 5. Some areas are viewed more than once during this 16-day repeat cycle because they lie in an area of overlap between adjacent orbits. The degree of overlap increases away from the equator.

As Landsats 1–3 moved southwards over the illuminated side of the Earth an instrument called the *Multispectral Scanner* (MSS) measured the reflectance of the land or water surface along scan lines extending 92.5 km on either side of the sub-satellite track (Figure 5.2). The width of one scan line is therefore 185 km. The measurements made by the MSS can be likened to the measurement

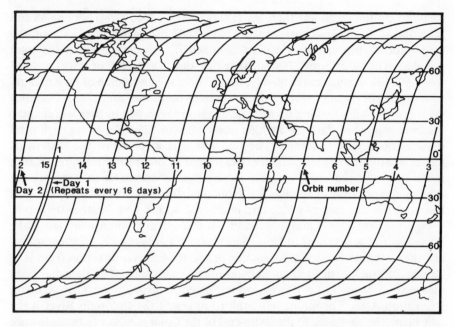

Figure 5.2   Orbit of Landsat-4 and -5 satellites. These satellites have a mean orbital height of 705 km and pass over all points between 82°N and 82°S at least once every 16 days. The equatorial crossing time on the southbound (descending) orbit is 0945 local sun time. The sub-satellite track is the line traced out on the Earth's surface by a point immediately below the satellite. The sub-satellite tracks for adjacent orbits become closer towards the Poles.

of light by a photographer's light-meter. Each Landsat MSS measurement is proportional to the amount of light reflected from a ground area (called a *pixel*, for picture element) with a side-length of 79 m. For technical reasons the values recorded on the ground are for areas measuring 57 m along the scan lines and 79 m between scan lines. The values recorded for each pixel are expressed on a scale of 0–63, giving a requirement of six bits of computer storage per pixel. On this scale, 0 signifies minimum detectable reflected light and 63 indicates maximum detectable reflected light. If the reflectance of sunlight from the ground or sea surface as recorded by Landsat's MSS were shown in photographic form then values of 0 would be seen as black and values of 63 as white with intermediate values being represented by shades of grey. Some Landsat data distribution centres rescale the pixel values onto a 0–127 or 0–255 range.

If 2342 scan lines from the MSS are placed together like the rows of a matrix then the distance from the first to the last scan line is just about 185 km. Each scan line extends 92.5 km to either side of the sub-satellite track. A Landsat MSS image is therefore made up of 2342 scan lines, giving the image a size of 185 × 185 km on the ground. Since pixels are recorded at 57 m intervals along each 185 km-long scan line there are (185 000/57 =) 3246 pixels per line. A full MSS image is made up of $2342 \times 3246 = 7\,602\,132$ pixels. An idea of the magnitude of this quantity can be obtained if it is compared to the total population of New Jersey (7 344 000) or Ecuador (7 814 000), or if it is realized that 7 602 138 seconds make 90 days and 10 hours. The volume of data making up a single MSS image places some restrictions on its use. Since the reflectance values for the area making up a single image are recorded in four wavebands, the total volume of data for a single scene is 30 408 528 bytes, since each pixel requires one byte of storage. The first image is a record of the pattern of reflectance of green light and images two to four record, respectiyely, reflectance patterns in the red and two near-infrared wavebands. These concepts are explained more fully in Section 5.3. Note that, for historical reasons, the four bands of the Landsat 1–3 MSS are numbered 4, 5, 6 and 7 rather than 1–4.

Figure 5.3(a) is a Landsat-3 band 7 (near-infrared) image of the Lake Powell/Glen Canyon area of Utah. The black, low-reflectance, areas are water, and the Colorado River is seen running in a south-westerly direction from Hite Crossing in the north to Glen Canyon in the south-west. The river has been dammed to form Lake Powell, which extends through the Glen Canyon Recreation Area. The major left-bank tributary of the Colorado River is the San Juan. The dark area to the south-east of the Colorado/San Juan confluence is Navajo Mountain, which rises to 10 388 feet. To the north-west of the Colorado River the Kaiparowits Plateau and the Escalante River are prominent. Figure 5.2(a) is a computer-generated digital image; some of the numbers forming this image (and representing ground reflectance values for individual pixels in the near-infrared waveband of the Landsat MSS) are listed in Figure 5.3(b). Figure 5.3(c) is a sketch map showing the locations of places mentioned in the text.

Although Landsats 4 and 5 have a lower orbital altitude (705 km) than Landsats 1–3, their orbits cover the same proportion of the Earth's surface – that between 82°N and S latitude. A MSS instrument like the one carried by Landsats 1–3 and described earlier is carried by the newer Landsats; in addition, they have a more modern scanner called the *Thematic Mapper* or TM. Unfortunately the electrical supply on Landsat-4 has partially failed, though the Landsat-5 TM has worked perfectly since launch. Landsat-4 is still in orbit and its MSS continues to function, but its TM instrument is not fully operational.

TM pixel values are recorded on a 0–255 scale rather than the 0–63 scale of the MSS. The TM operates on the same principles as the MSS, but it has a ground resolution or pixel size of 30×30 m rather than 57×79 m, and it records ground reflection in six, not four, wavebands. A seventh band covers the spectral region in which heat is emitted by the Earth's surface. The ground resolution in this "thermal" band is 120 m. The Landsat 4 and 5 MSS bands are numbered 1–4 and the TM bands 1–7. Figure 5.4 is a near-infrared TM image of an area to the south of the city of Topeka in Kansas. The runways of the city airport

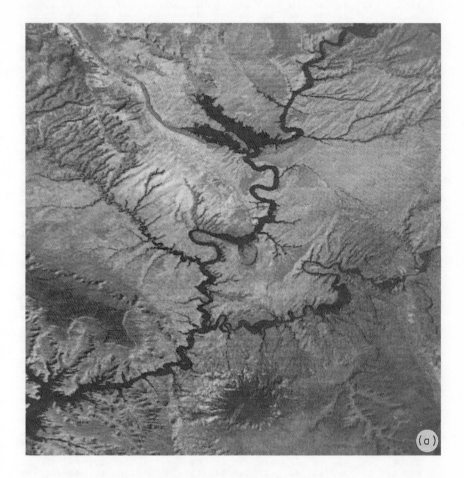

Figure 5.3 **(a)** Landsat-3 band 7 (near-infrared) image of the Lake Powell/Glen Canyon area of Arizona and Utah, United States. The main features are shown in Figure 5.3(c).

can be seen on the extreme centre right of the image, while the route of Interstate Highway 135 (Topeka to Wichita) runs diagonally across the area from lower left to top right. The black patch in the top left corner of the image area is a small reservoir. Brush-filled creeks (showing as dark streaks) drain the region towards the Wakarusa River in the south-east, and the regular criss-cross pattern of tracks and field boundaries stands out clearly.

Both MSS and TM images are composed of individual pixels. It was mentioned earlier that a remotely-sensed image is made up of a number of scan lines and that each scan line contains a large number of pixel values. Each pixel has an

```
44 35 42 40 31 33 35 40 33 44 42 46 37 37 40 42 ·42
45 34 38 29 31 40 45 40 36 45 43 43 40 36 43 43 45
43 41 38 27 25 52 43 36 29 43 36 47 34 43 34 45 43
42 42 33 29 35 46 46 35 46 42 42 44 37 35 35 40 37
40 45 40 47 49 54 36 38 40 43 27 43 43 34 47 38 22
41 34 22 34 32 38 34 34 36 45 43 43 41 29 34 16 36
31 13 33 53 33 37 40 29 31 20 37 44 44 33 44 51 49
40 15 52 43 38 31 34 45 22 40 31 36 45 40 27 43 36
32 50 43 52 41 38 50 22 32 29 22 50 36 36 34 34 36
11 57 40 42 40 37 37 37 33  7 37 46 42 37 35 31 33
56 54 40 43 43 36 31 20 18 34 61 43 38 47 36 34 36
52 47 43 36 32 34 18 43 50 41 59 57 41 43 32 32 43
33 46 37 33 18  7 53 44 40 46 51 42 44 40 40 31 35
47 36 34 20 38 31 58 36 45 49 45 43 40 45 38 43 36
41  7 34 34 43 59 38 63 63 50 52 47 36 43 36 36 38
42 37 40 42 51 35 64 51 46 51 40 35 26 37 42 37 37
49 27 45 38 56 38 54 38 49 49 40 36  9 58 47 34 36
50 29 47 50 32 50 52 34 38 43 34  9 57 54 38 36 34
29 49 44 40 40 53 49 35 35 29  7 53 55 40 40 40 46
45 52 68 29 70 38 56 18  6 36 54 56 65 36 47 38 36
50 68 36 41 61 47 25 45 54 47 61 29 36 34  7 47 52
51 71 37 62 62 33 18 44 55 49 37 37 31  4 53 55 40
54 70 27 68 61  4 68 27 61 40 40 27 13 47 56 58 34
61 41 34 59 43 36 36 61 50 38  7 54 47 79 57 32 34
53 57 46 53 49 51 33 66 44  7 66 55 22 24 40 37 53
40 49 45 54 61 13 49 63  4 61 56  6 38 49 63 49 56
34 43 57 59 47 47 57 47 57 52 13 36 70 68 52 52 54
49 55 49 66  9 62 53 33 24 24 42 46 57 55 62 79 46
52 31 68 52 54 56  6 36 49 65 47 45 68 65 47 40 43
34 50 66 54 63 38 25 59 59 45 57 72 36 34 45 43 68
33 49 51 40  9 33 53 53 57 64 40 49 42 46 31 49 62
34 63 11 29 31 52 63 49 61 38 72 45 52 34 49 52 61
68 27 32 41 57 52 63 43 57 57 27 41 22 52 52 59 47
37 35 53 44 49 73 49 62 33 31 26 53 33 42 53 60 44
 9 47 61 61 61 45 25 18 45 58 38 56 22 40 61 18 47
22 59 38 52 41 43 45 63 47 43 57 61 45 32 70 63 52
44 66 37 18 42 49 42 57 31 51 71 60 51 57 33 49 40
58 31 40 47 49 47 43 68 36 56 63 74 61 52 40 40 40
47 34 50 57 50 50 32 54 50 50 57 61 45 45 36 57 36
33 26 35 53 49 49 49 53 55 66 37 42 62 46 60 24 49
29 52 47 54 49 18 49 58 31 27 31 36 43 47 31 56 65
50 50 45 61 52 38 68 36 32 50 52 50 36 36 52 63 54
42 49 60 57 42 57 15 42 46 44 57 40 42 49 53 66 37
45 58 54 54 65 38 38 54 54 54 38 54 52 54 43 49 38
52 61 47 52 18 41 41 63 41 54 32 61 36 47 63 36 54
55 55 57 26 44 44 44 51 44 49 35 37 60 57 44 51 37
52 43 52 43 56 52 54 54 29 54 40 58 49 49 47 36 63
50 45 20 47 52 45 47 54 50 43 52 52 38 57 45 57 47
46 51 22 49 29 51 35 35 35 42 49 51 42 62 42 55 53
(b) 54 54 27 45 34 38 54 34 52 47 34 52 56 36 61 54 54
```

Figure 5.3 (*cont.*)   (b) Section of Figure 5.3(a) in digital form. Each number measures the reflectance of a small ground area (a pixel) with dimensions 79 × 57 m. The reflectance values are expressed on a 0–127 scale. These digital reflectance values are converted to a viewable image by the computer system shown in Figure 5.10.

associated numerical reflectance value measured on an ordinal scale (0–63 for the MSS and 0–255 for the TM). Because Landsat MSS and TM images are numerical in nature they are called *digital images*. They can be processed, manipulated and displayed by computer. This chapter contains an introduction

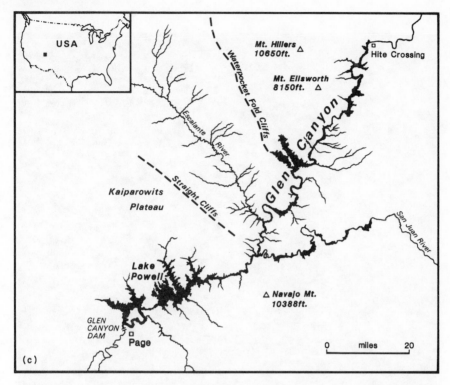

Figure 5.3 (*cont.*) (**c**) Sketch map of the area covered by the image shown in Figure 5.3(a).

to methods of digital image processing applied to Landsat image data. First of all, however, the physical principles on which remote sensing is based are reviewed.

5.3 PHYSICAL BASIS OF REMOTE SENSING

Landsat MSS and TM instruments measure the reflectance of sunlight (or, in the case of the TM thermal infrared band, conventionally numbered 6, the emittance of heat) by the Earth at each of a very large number of pixel locations. Sunlight and heat are two kinds of electromagnetic energy and remote sensing is concerned with the interaction between electromagnetic energy and Earth surface materials such as vegetable matter, water, soil and rock.

Electromagnetic energy can be thought of as moving in a wave-like pattern at the speed of light. The crest-to-crest distance between adjacent waves is called the wavelength of the energy (Figure 5.5). The wavelength is measured in

Figure 5.4   Near-infrared image of the area around Topeka, Kansas, produced from
data recorded by the Landsat-4 Thematic Mapper.

fractions or multiples of a metre. Short wavelengths are measured in millionths of a metre (10^{-6} m or micrometre, μm) or even billionths of a metre (10^{-9} m or nanometre, nm). To give an idea of the magnitude of a nanometre, consider that light travels one foot in one nanosecond (10^{-9} s). Longer wavelengths are measured in hundredths of a metre (centimetre, cm), metres (m) or in thousands of metres (kilometre, km). This range of wavelengths – from billionths of metres to kilometres – is called the electro-magnetic spectrum, or simply the spectrum. For ease of description it is divided up into groups of wavelengths termed wavebands or spectral bands (Figure 5.6). Only certain parts of the spectrum are used in remote sensing

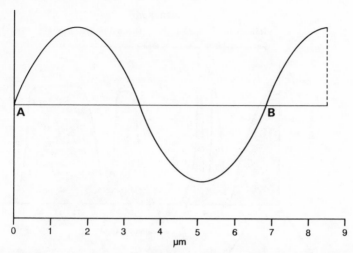

Figure 5.5   The wavelength of a symmetric curve is the distance between corresponding points on adjacent waves. The amplitude of the curve is its maximum height above its average level. The curve shown has a wavelength of approximately 6.6 μm.

because radiation at some wavelengths is absorbed by gases present in the atmosphere, particularly ozone and water vapour.

The shortest wavelengths are grouped into wavebands called gamma rays, X-rays and ultraviolet (UV) radiation. Much of the energy at these wavelengths is absorbed by the atmosphere, which means that a remote sensing instrument such as the MSS or TM would, if it operated in one of these bands, be unable to see through the atmosphere and consequently would be unable to obtain any information about the surface of the Earth.

Visible light has a longer wavelength than UV radiation. It occupies that part of the spectrum between 0.4 and 0.7 μm. The 0.4–0.5 μm band is perceived by our eyes as blue light, whereas the 0.5–0.6 μm region appears green and the 0.6–0.7 μm band is seen as red light. The diameter of a full stop (period) character in ordinary type is 0.06 cm, so a ray of red light would have a wavelength 92 000 times smaller than the diameter of a full stop.

Beyond the red end of the visible spectrum is the infrared or IR. The IR waveband is not homogeneous. The short-wave infrared (SWIR) that lies closest to the red end of the visible spectrum behaves like visible light (except that our eyes cannot detect it). Photographs using SWIR radiation rather than visible light can be taken with an ordinary camera and special film. IR radiation with wavelengths between 5 and 15 μm is sensed as heat and so it is called the thermal infrared region. The Earth emits heat in the thermal IR band in a waveband centred at 9.6 μm. This heat can be detected above the atmosphere by thermal sensors which form part of Landsat's TM and by thermal imaging instruments carried by meteorological satellites such as NOAA and Meteosat.

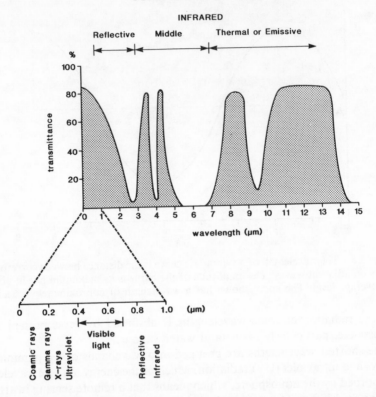

Figure 5.6   The electromagnetic spectrum from 0 to 15 μm. The vertical axis shows the proportion of energy transmitted through the Earth's atmosphere. Regions of high transmittance are called "atmospheric windows". These are the regions of the spectrum that are available for remote sensing.

Microwaves have wavelengths much longer than the thermal IR. They are really very short-wavelength radio waves (0.3 cm–3 m). Radar (Radio Detection and Ranging) uses the 0.8 cm–1 m waveband. Electromagnetic energy in this region has two very valuable properties. It can penetrate fog and cloud, and can also be used at night. Radar images measure the roughness (rather than the colour or the temperature) of the target surface. No civilian radar satellite is in orbit at the moment, though the US Seasat produced much valuable imagery during its short lifetime in 1978. The European Space Agency and the Canadian Government are both planning to launch radar satellites in the early to mid-1990s.

The portion of the electromagnetic spectrum with wavelengths longer than 3 m is used for radio and TV communications. FM radio uses the shortest wavelengths, while AM stations use the 20 m to 2 km band. Short-wave radio uses the 20–50 m band, medium wave covers the 190–550 m band while longwave uses the 1000–2000 m region.

Only a part of the electromagnetic spectrum can be used for remote sensing for, as noted earlier, energy in some spectral bands is absorbed or scattered by the atmosphere. All energy with a wavelength less than 300 nm is absorbed, and there are a number of absorption bands in the IR region between 1.1 and 2.5 μm. Apart from the 3–5 μm and 8–14 μm bands, infrared energy is absorbed. The regions of the spectrum that are not affected by absorption are called atmospheric windows (Figure 5.6). Energy with wavelengths within these window regions is not absorbed but may be subject to a process called *scattering* which deflects or redirects the energy. Scattering gets more intense as wavelength diminishes, so blue light in the visible spectrum is scattered more than red light because the wavelength of blue light is less than that of red light. The blue component of incoming solar radiation is scattered so severely that it appears to our eyes to be coming from the entire sky. That is why the sky looks blue. The Moon has no atmosphere, and photographs taken on the Moon's surface show a black sky.

The degree of scattering is affected by the distance traversed through the atmosphere, called the atmospheric path length. If you have travelled in an aeroplane at 35 000 feet or so then you will have noticed that the sky looks a much darker blue than at sea-level. This is because there is less scattering of blue light as the atmospheric path length and consequently the degree of scattering of the incoming radiation is reduced. For the same reason, the sun appears to be whiter and less orange-coloured as the observer's altitude increases; this is because a greater proportion of the sunlight comes directly to the observer's eye. Figure 5.7 is a schematic representation of the path of electromagnetic energy in the visible spectrum as it travels from the sun to the Earth and back again towards a sensor mounted on an orbiting satellite. The paths of waves representing energy prone to scattering (that is, the shorter wavelengths) as it travels from sun to Earth are shown. To the sensor it appears that all the energy has been reflected from point P on the ground whereas, in fact, it has not, because some has been scattered within the atmosphere and has never reached the ground at all. This component is called the atmospheric path radiance. Another component has been reflected from points other than P. Only a part of the received signal represents energy reflected from point P. The effects of scattering diminish the usefulness of shorter wavelengths for remote sensing.

5.4   GEOGRAPHICAL SIGNIFICANCE OF LANDSAT MSS AND TM DATA

You may well be wondering about the relevance of this technical detail on orbits, sensors and electromagnetic energy to geography. However, it is necessary to understand the principles of remote sensing in order to make intelligent and informed use of remotely-sensed data. Images produced from these data have the following geographically-significant properties:

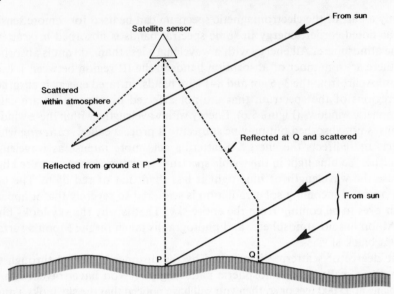

Figure 5.7   Part of the short wavelength energy from the sun is scattered at S2 into the field of view of the sensor. Some energy reaches point P on the Earth's surface and a proportion is reflected into the field of view of the sensor. Some energy from point Q is also detected by the sensor as if it had originated at P, due to scattering at S1. Thus, although to the sensor all the captured energy appears to come from point P on the ground, only a proportion actually does so.

1. They provide coverage of the Earth between 82°N and S latitude, providing information about areas to which access on the ground might be difficult or expensive.
2. The coverage is repetitive (every 18 days for Landsats 1–3, every 16 days for Landsats 4 and 5) so phenomena which change through time can be monitored. Cloud-cover problems will diminish the number of useful images from the 20–22 per year that are theoretically possible, but even so the monitoring of surface phenomena that change during the year (such as agricultural crops and natural vegetation) or which change over the years (for example the extent of the built-up areas of cities or the extent of forest cover) is possible.
3. The TM and MSS produce images from a considerable height above the Earth's surface so panoramic distortion (familiar to users of low-altitude aerial photography) is minimal. The MSS "sees" through an angle of only 11° so the distance from the sensor to the ground at any point on the image is not too variable.
4. TM and MSS images are multispectral. The TM produces images in seven spectral bands, the MSS in four. This makes possible the identification of ground surface features from their spectral characteristics.

5. The images that the Landsat MSS and TM instruments provide are digital or numerical in nature, and can therefore be processed by computer. Indeed, without the use of a computer it would not be possible to use much remotely-sensed imagery at all. Apart from some photographs taken by astronauts using hand-held cameras, and some experimental photographs taken by cameras carried by the Space Shuttle, all remotely-sensed imagery from satellites is digital in nature. A computer is needed to convert these images from numerical to viewable form, either as a picture on a TV monitor or as a conventional photograph.

The multispectral nature of Landsat TM and MSS data requires further comment. A green object is perceived as green because it is more reflective in the green region of the visible spectrum than in the blue or red regions. A plot of the spectral reflectivity curve for a bright green target is shown in Figure 5.8. Both the TM and MSS operate in spectral regions beyond the visible. Although our eyes cannot detect variations in reflectance in the near-infrared band it is possible to measure such variations using appropriate instruments. Figure 5.9(a) shows the spectral reflectance curves for healthy vegetation (solid line) and for diseased vegetation (dashed line) in the 0.4–1.1 μm region of the spectrum. The most noticeable aspect of the reflectance curve (or spectral signature) of vigorous vegetation is the sharp rise in reflectivity at 0.7 μm; the green peak (0.5–0.6 μm) is relatively minor. The curve for diseased vegetation shows that the near-infrared is considerably reduced. Other targets are characterized by specific spectral reflectance properties. For example, water has a generally low reflectance in the visible spectrum, the proportion of reflected light being related to the presence and amount of suspended sediment and dissolved organic material. In the near-infrared waveband the reflectivity of water drops almost to zero. If the spectral reflectance characteristics of the

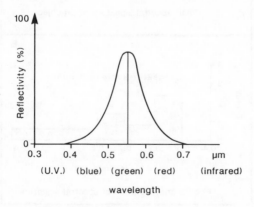

Figure 5.8   Spectral reflectance curve for a bright green target.

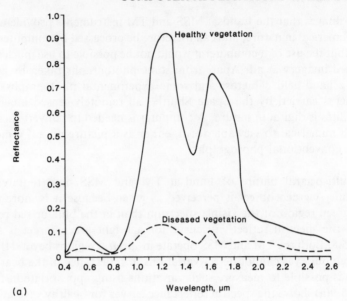

(a)

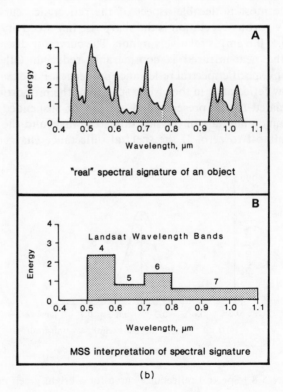

"real" spectral signature of an object

MSS interpretation of spectral signature

(b)

different land-cover types in the area covered by an image are known, it is possible to identify each pixel in the image in terms of one or other of these classes. However, the broadness of the MSS bands (Figure 5.8(b)) precludes such precise identification. It should also be remembered that the spectral reflectance curve of living vegetation will change continuously throughout the growing season.

5.5 COMPUTER SYSTEMS FOR LANDSAT DATA

Of all the data types described in this book, Landsat data are the only kind that are interpreted and used in the form of pictures or images. Any computer system used for Landsat data processing must be capable of displaying images, preferably in colour. In this section the equipment needed to allow the computer to display colour images on a TV monitor is described.

Landsat TM and MSS data are provided in computer-readable form on magnetic tapes called CCTs (Computer Compatible Tapes). The data stored on these tapes are read by the computer and placed in a file (or files) on disc (Chapter 1). After any processing of the kind described later in this chapter the data are transferred to one of three memory banks, each of which contains a number of parallel bit-planes; each plane is capable of storing one bit (0 or 1) at each pixel position (Figure 5.10). The pixels within each memory bank are arranged in row and column format, and a common arrangement is to have 512 rows (or rasters) of 512 pixels in each of the memory banks. The three memory banks represent the red, green and blue components of the image seen on the television monitor. Generally, eight bit-planes make up one memory bank so that a value between 0 and 255 inclusive (00000000 to 11111111 in base 2 notation, described in Chapter 1) can be stored at each pixel position for the three primary colours of red, green and blue. Figure 5.11 shows how the value stored at one pixel position in a memory bank is mapped to the corresponding position on the TV monitor display. Since in the example the memory bank is composed of one bit-plane, the resulting image shown on the TV monitor can have one of only two values at each pixel point: the value 0 (meaning "black") and the value 1 (meaning "white"). No intermediate shades of grey are possible. Figure 2.2 shows the letter I represented in raster format with one

---

Figure 5.9 *opposite* (a) Spectral reflectance curves for healthy and diseased vegetation. Note the minor peak at 0.55 μm (in the green region of the visible spectrum) and the sharp rise in reflectance at about 0.8 μm. The dips in the reflectance curve for healthy vegetation at 0.4 and 0.6–0.8 μm are due to absorption by the growing vegetation to provide energy for the process of photosynthesis. (b) The actual reflectance spectrum of an object (upper diagram) and the spectrum recorded by a broad-band sensor such as Landsat's Multispectral Scanner.

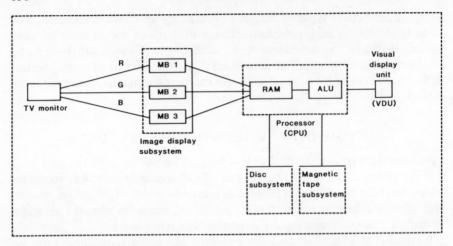

Figure 5.10   Schematic representation of a dedicated image-processing computer system. Remotely-sensed images are read from magnetic tapes and stored as disc files. These data are transferred, under the control of a program running on the processor, to an image display sub-system which contains three memory banks. These memory banks store the red, green and blue components respectively of a false-colour composite image. The digital images held in the memory banks are converted to television signals which are displayed on the monitor. The operator issues commands via the keyboard associated with the visual display terminal. The processor box holds the arithmetic and logical unit (ALU) and the random-access memory (RAM).

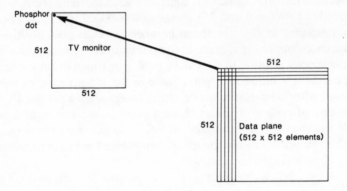

Figure 5.11   The numbers stored in the memory banks shown in Figure 5.9 represent the intensity of the colour displayed at the equivalent point on the TV monitor screen. In an operational system the memory is capable of storing a number in the range 0–255 in each cell of the 512×512 array. A value of 255 would result in the corresponding phosphor dot on the screen receiving maximum illumination while it would be at half brightness if the value were 127.

bit per pixel. Recall, however, that dedicated image processing computers such as that shown in Figure 5.11 have several bit-planes in each of three memory banks. Eight bit-planes per memory bank allows 256 levels of the associated primary colour to be represented.

The relationship between the number of shades of a colour (or shades of grey if the image is monochrome) and the number of bit-planes in the memory bank can be illustrated by a simple example. A single bit-plane can hold only one binary digit in each position. Two planes can store the binary numbers 00, 01, 10 and 11 (decimal 0, 1, 2, 3) and so could represent black (0) and white (3) with two intermediate shades of grey. In general, n bit-planes can hold 2^n different levels of grey (including black and white) at each pixel point. A group of data planes makes up a memory bank. Since Landsat TM data are supplied on a 256-level scale it would seem sensible to have a memory bank comprising eight data planes, since 2^8 equals 256. It is possible to use less than eight data planes and still get good results. The Landsat MSS image shown in Figure 5.3(a) was photographed from a monitor screen in the Remote Sensing Unit at Nottingham University using six data planes per memory bank (implying 64 levels of colour per memory bank).

There are two ways of generating a colour representation of an image. One is to allocate a particular colour to a number of contiguous levels in the image so that, for example, the levels 0–32 might be displayed in dark blue, 33–67 in dark green, 68–87 in cyan, and so on. This method is called *density slicing* and it was the technique used in the generation of the colour image shown on the front cover of this book. Density slicing is normally applied when only a single-band image of an area is available. When multispectral imagery is available, such as from Landsat's MSS or TM, three bands are selected for display. The data for band 1 are placed in memory bank 1 which provides the red input to the TV monitor display. Band 2 is stored in memory bank 2 (green) and band 3 in memory bank 3 (blue). The resulting picture on the monitor is called a *false-colour composite* image. It is a composite of three separate bands and is false (rather than natural) colour because the three selected spectral bands do not usually represent reflection in the visible red, green and blue regions of the spectrum. Thus, a Landsat 1–3 MSS false-colour composite image is generally made up from band 7 (shortwave infrared) displayed as red, band 5 (red) displayed in green and band 4 (green) displayed in blue. From the earlier discussion of spectral signatures it is easy to understand why vigorous vegetation appears red in this type of display.

Colours other than pure red, green and blue are formed by the combination of the three primary colours. For instance, the colour yellow is formed by the addition of red and green. A Landsat MSS pixel that had equal reflectance in bands 7 and 5 and zero reflectance in band 4 would appear as a yellow point on the display if the false-colour compositing method described in the preceding paragraph were used. The brightness of the yellow would be a function of the magnitude of the reflectance. Cyan is formed by the addition of green and blue, while magenta (pink) is the result of combining red and blue. Virtually any colour can be formed by mixing red, green and blue light in different amounts and with different levels or magnitudes. Table 5.1 shows some possible combinations. Note that if there are eight data planes in each memory bank (giving 256 levels per colour) then the number of combinations is 256^3 or 16 777 216. The human eye cannot distinguish such a large number of colours and no Landsat MSS image would need to use them all. However, should a particular colour be needed it will be available.

Many image-processing computer systems can perform a number of straightforward operations on the displayed image. One useful operation is zooming or magnifying a portion of the image so that it fills the screen. Zooming is carried out by repeating pixel values, so that zooming by a factor of four implies that each pixel value in the memory bank or banks is repeated four times, and each line of the image is also repeated four times. Figure 5.12 shows a part of the image in Figure 5.3(a) magnified by a factor of 4. The area shown is the junction of the San Juan and Colorado rivers in Utah. Note the prominent abandoned meander on the Colorado River north of the confluence. It can also be seen, though less clearly, on Figure 5.3(a). The individual pixels of which the image is formed are clearly apparent in Figure 5.12.

Table 5.1   Digital intensity values on 0–255 scale (eight bits) for three memory banks representing, respectively, red, green and blue

Memory bank 1 (red)	Memory bank 2 (green)	Memory bank 3 (blue)	Perceived colour
255	0	0	maximum red
0	255	0	maximum green
0	0	255	maximum blue
255	255	0	maximum yellow
127	127	0	mid-yellow
0	255	255	maximum cyan
0	127	127	mid-cyan
255	0	255	maximum magenta
255	255	255	white
0	0	0	black
127	127	127	mid-grey

The colour seen on a monitor screen is named in the right-most column. Over 16 million different colours are possible using eight-bit representation for the three additive primary colours since any value in the range 0–255 inclusive can be selected for any of the three memory banks.

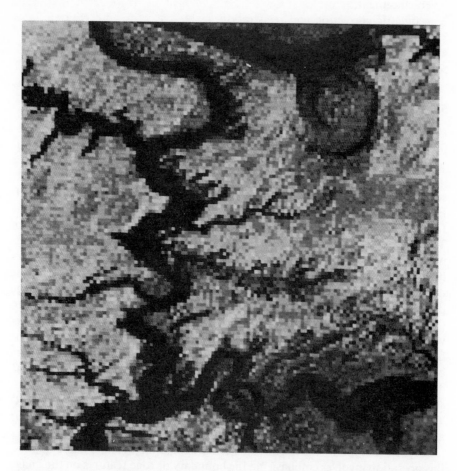

Figure 5.12   Zoomed portion of the image shown in Figure 5.3(a). The confluence of the San Juan and Colorado rivers can be seen in the lower centre-left of the image. At this magnification ($\times 4$) the individual pixels making up the image become visible.

5.6   IMAGE PROCESSING TECHNIQUES

Methods of displaying and magnifying digital images were discussed in the preceding section. Since a digital image is simply an array of numbers it can be manipulated by computer; the type of manipulation depends on the aims of the user. In this section a number of techniques of digital image processing which can be applied to Landsat TM and MSS images are described. These techniques can be considered to alter in some predetermined fashion the way in which the image is presented to the viewer. They are described in the following paragraphs under five headings:

1. enhancement,
2. geometric correction and registration,
3. noise suppression and filtering,
4. classification, and
5. transformation.

5.6.1 Enhancement operations

Enhancement operations are those which bring out detail in an image, either by improving the contrast or by emphasizing edges. The latter operation is sometimes called *sharpening*. The raw data received from Landsat often use only a portion of the 0–255 range available in an eight-plane memory bank, so that the image appearing on the screen is either too dark or too light. Enhancement techniques spread the range of pixel values so that they cover the full dynamic range of the image display system. For instance, a typical Landsat TM band 4 image might have pixel values in the range 25–90. If this image were displayed using the methods described earlier it would lack very dark values (0–24) as well as medium to bright values (91–255) and would cover only the dark to medium grey range. The computer could be used to rescale the image pixel values so that the lowest value in the image (25) was interpolated onto the lowest value that the display system could accommodate (0) while the highest value in the raw image (90) was mapped to the maximum value that could be held at a pixel point in the image memory bank (255). The intermediate values between 25 and 90 would be interpolated onto the 0–255 range of the display device. This technique is called *contrast stretching*. Figure 5.13 is a contrast-stretched Landsat MSS band 7 image of the Leningrad region in the USSR. The original image lacked any detail yet on Figure 5.13 the port installations and the delta of the Neva River are clearly brought out.

Another method of enhancement is called *sharpening*. As the name implies, this operation involves emphasizing the boundary or edge features on the image so that it looks less blurred. An edge or boundary can be thought of as a point or pixel across which the reflectance value changes considerably. Figure 5.14 was produced by (a) moving across the image on a pixel-by-pixel basis and computing the difference between a given pixel and its neighbours to the top, bottom, left and right, and (b) subtracting four times this difference from the central pixel value. The effect is to highlight edges and boundaries such as railway tracks, roads and streets. It also has the undesirable effect of enhancing the horizontal scan line pattern, particularly over low-reflectance areas such as the water of the Gulf of Finland.

5.6.2 Geometric correction and registration

Geometric correction and registration are necessary if Landsat imagery is to

Figure 5.13   Landsat MSS band 7 (near-infrared) image of Leningrad and the upper
Gulf of Finland.

be used for cartographic purposes, or is being used in conjunction with map
data (for example in a Geographical Information System, Chapter 7). A raw
Landsat TM or MSS image does not conform to any standard map projection.
First of all, in the case of the MSS, the pixels are not square. They each represent
an area measuring 79×57 m on the ground, giving the image a compressed
appearance when it is displayed on a TV monitor. Secondly, Landsat's orbit
does not follow a north–south track. Thus, the scan lines (which are
perpendicular to the satellite heading, or flight direction) do not run in an east–
west direction. At the equator, Landsat's heading is 189° (9° west of south).
The heading increases with latitude until at 82° latitude the satellite is going

Figure 5.14   Same image as shown in Figure 5.13 after edge-enhancement procedure has been applied. Although the image is sharper, the enhancement process has brought out a horizontal banding pattern on the image.

westwards (Figure 5.1). Thirdly, Landsat's orbit is not perfectly stable. Its height varies by up to 30 km and its attitude can change slightly, implying that the pointing direction of the sensor is not always truly vertical. If image data are to be correlated with map data then these properties of the image data must be removed and the pixel values forming the image must be re-expressed in terms of a map-compatible coordinate system at the required scale.

The operations needed to carry out this procedure can be divided into two parts. Firstly, control points are located on map and image. These points should be easily-recognizable and accurately-measurable points such as runway

Figure 5.15 Point locations on an image can be found using a cross-wire cursor. In this photograph the cursor has been placed on the runway of Lakenheath air force base in eastern England. Collection of ground control point coordinates is the first stage in the geometric correction of an image (see text for discussion).

intersections at airports, railway or road junctions, isolated buildings, or prominent natural features. The location coordinates of suitable points on the image can be found accurately if a zoom function is available, and if a cross-wire cursor can be positioned over the point. Figure 5.15 shows a control point being located on a magnified Landsat MSS image of Lakenheath USAF base in eastern England. Around 30 or 40 such ground control points (gcp's) are needed for each full 185×185 km image. For each point the map coordinates and the image column and row coordinates are recorded, and from these a mathematical transform is computed. This transform is essentially an equation which will convert from image to map coordinates and vice-versa.

The second stage of the geometric correction procedure involves the relocation of image pixel values to new positions which are required to achieve conformity between map and image. This procedure is essentially a two-dimensional interpolation, and is known as *resampling*. The map coordinates of the required pixel position are converted to image coordinates using the transform computed at stage one, and the pixel value on the raw image that is nearest to the computed image coordinate position is selected, as shown in Figure 5.16. This process is repeated for each required pixel position in the transformed image.

The same procedures (locating ground control points, computing a map-to-image and image-to-map transform, and resampling to generate a corrected output image) can be used to register or overlay two images of the same area taken at different dates. Registered images are used to assess the degree of change that has occurred during the time-period represented by the two images.

Both geometric correction and image registration involve lengthy and time-consuming operations, taking several hours on a minicomputer. Because the resampling operation is both repetitive and independent (that is, two or more resampling operations could be carried out simultaneously – one does not rely upon the other) it does not need to be performed sequentially. Nearly all present-day computers are sequential (serial) machines which perform only one operation at a time. A parallel computer is described in Section 1.8.3, and it is easy

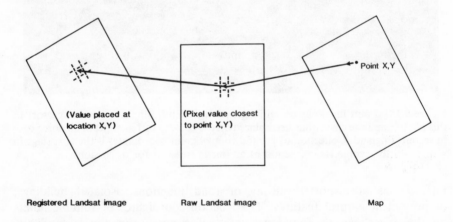

Registered Landsat image   Raw Landsat image   Map

Figure 5.16 The geometric correction of a remotely-sensed image involves the computation of equations which allow map coordinates to be converted to image coordinates, and vice-versa. Given a raw Landsat image (centre) and a map of the area of interest (left) a corrected image, registered to the map, is required. This registered image is derived by taking the x,y map coordinates of points corresponding to pixel centres on the registered image and locating the point on the raw image. The pixel value in the raw image closest to the computed point is placed at location x,y in the registered image.

to appreciate how the use of such a machine could speed up an important operation on digital remotely-sensed images.

5.6.3 Noise suppression and filtering

Noise suppression and filtering involve the amplification or the removal of some of the components of the image data. Noise in this context means the unwanted products of the imaging system. Landsat images are scanned on a line-by-line basis, and this line-scanning process often leads to the production of horizontal striping effects on the resulting image. These stripes are an example of noise, which can be defined as random or systematic effects superimposed upon the image due to the characteristics of the imaging system and the electronic circuits used in storing, transmitting, recording or copying the image. The edge-enhanced Leningrad MSS image (Figure 5.14) illustrates one way in which a particular component of the image (the boundaries or edges) are amplified. The method used to achieve this could be called a *filtering* technique for, by analogy with the process of filtering as used in the chemistry laboratory, a particular part of a mixture (the edges or boundaries) was isolated and then added back to the original image to give Figure 5.14. Thus, an image can be considered to be a mixture of components, each component representing variation at a particular scale. It is convenient to distinguish between two scale components of an image. One is the overall low-frequency background pattern of dark and light, and the other is the rapidly-changing local variation, or the high-frequency component, which is superimposed upon the low-frequency background to produce the observed image. Pixel value at any point on the image can therefore be considered to consist of

(a) a contribution from the background or regional pattern, and
(b) local variability or detail which is superimposed upon this background pattern.

This way of modelling the scale components of an image should be compared with the Trend Surface model described in Section 3.5.3.

Filters can be designed to extract either the background pattern or the pattern of local variation. These filters are called *low-pass* and *high-pass* respectively. An example of the use of a high-pass filter was given above in respect of Figure 5.14. A high-pass filter was used to isolate the local (high-frequency) variation which was then added back to the image so that the local component was effectively doubled, thus amplifying or exaggerating its importance. This operation is termed sharpening.

A low-pass filter does the opposite of sharpening. It smooths or blurs the image so that local deviations from the overall trend are removed. The local variation may not represent information at all for, as Figure 5.14 clearly shows, some high-frequency noise in the form of horizontal scan line patterns may be present on an image, distracting the user and rendering him or her less capable of recognizing underlying trends. The effect of removing noise is to clean up the image while the effect of removing local variability is to smooth the boundaries and emphasize the regional pattern present in the image. Figure 5.17(a) is a Landsat MSS image of a coastal area of eastern England called The Wash. The tidal range in this estuary is large (7 m at spring tides) and large sandbanks are exposed at low tide. The suspended sediment from rivers discharging into the The Wash shows up in grey, contrasting with the black of the deeper, clearer water further offshore.

On Figure 5.17(a) the effect of horizontal scan line banding is apparent, and local variability reduces the interpretability of the image. Figure 5.17(b) was produced by replacing each pixel value in Figure 5.17(a) by the average of the

Figure 5.17   **(a)** Landsat MSS image of The Wash, eastern England. A horizontal striping pattern distracts the eye and disguises the pattern of light and dark.

pixel values in a 7 pixel by 5 line box centred on that pixel; this is called a *moving average* filter. A greater degree of smoothing could have been produced by increasing the size of the box, but this implies the loss of more detail. The balance between the intensity of smoothing and loss of detail can only be worked out by trial and error; the effect of the 7×5 moving average filter on the visual appearance of Figure 5.17(a) is sufficient to produce a more usable product which reveals the location and shape of the offshore sandbanks and the disposition of the suspended sediment plumes from the rivers discharging into The Wash. Students of English history might like to note that the contents of

Figure 5.17 (*cont.*) (**b**) A moving-average filter removes the striping from Figure 5.17(a) and makes the underlying pattern clear. The dark area in the top right is deep, clear water. Sandbanks and coastal marshes are now clear, as are the variations in the sediment load of the estuarine waters. However, the field patterns over the land area in the upper left of the image are blurred by the smoothing operation.

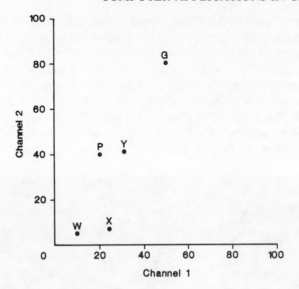

Figure 5.18   Points W, P and G represent the average reflectance values for samples of water, pine forest and grassland respectively on two channels of a multispectral image. Points X and Y represent unknown pixels. Since X is closest to W then it is allocated to class "water". Similarly, the pixel corresponding to Y is allocated to class "pine forest" since it is closest to point P.

the English treasury were lost in the thirteenth century by King John in the marshy area shown at the bottom left corner of the image. They have never been recovered.

5.6.4   Classification

Classification of the pixels forming a Landsat MSS or TM image means giving each pixel a label to associate it with a land-cover class such as "wheat", "forest" or "sandy desert". The four MSS bands provide four measures of surface reflectance which can be used to characterize different land-cover classes, as noted in Section 5.3. A classified image can subsequently be geometrically corrected (Section 5.6.2) to produce a thematic map, which can be used alongside other conventional maps showing, for example, soil, geological and climatic boundaries.

In order to classify the pixels it is necessary to know in advance the number and spectral characteristics of the land-cover classes that are present in the area covered by the image. Some methods of achieving this knowledge automatically are available but they are less reliable than methods based on field-work or the study of air photographs, maps and documentary records for the area concerned. A simple (yet still surprisingly efficient) method requires that the number of land-cover classes is known, and that for each class an estimate of the average reflectance in each Landsat MSS or TM band is available.

A simple example using only two bands is used to illustrate the method. Assume that, in terms of the two bands, the average reflectance of water, pine forest and grassland are respectively (10,5), (20,40) and (50,80). Figure 5.18 shows these points plotted on a graph. The point marked X is an unknown pixel which is to be labelled as water, pine forest or grassland. To do this the distance on the graph from X to each of the three points (water, grassland and pine forest) is calculated using Pythagoras's theorem, and pixel X is allocated to the class for which this distance is the shortest. In this instance, X is closest to the average value for water (point W) so it is labelled as a water pixel. Point Y, on the other hand, is closest to P, the average value for pine forest, so it is allocated to that class. This process is repeated for every pixel in the image so that, when the operation is complete, each pixel is labelled "W", "P" or "G" (in practice a numerical label would be used). This classified image could be displayed by assigning the colour blue to all pixels labelled "W", the colour dark green to all pixels labelled "P" and the colour light green to all pixels labelled "G". Figure 5.19 shows a classified image of an area of the county of Norfolk in eastern England. Shades of grey have been used to represent the different land-cover classes. Black areas are coniferous forest, which grows on the sandy soils of the Breckland, which is an area of glacial outwash material, while the dark grey, mid-grey and light grey regions represent different agricultural crop types. The mid-grey is permanent grassland while the light areas are intensively-cultivated fields of peas, beans and other cash crops. The different shades of grey were chosen subjectively and have no intrinsic meaning. Usually, colours rather than shades of grey would be used.

Classification techniques such as this have been used to produce thematic maps of land-cover types, of rock types in semi-arid and arid regions and of the extent of urban areas. One point that must be borne in mind is that such maps cannot be produced using Landsat MSS or TM data alone. Some knowledge of ground conditions is necessary. There is a growing tendency to use Landsat images as one of many layers of information within a Geographical Information System (Chapter 7).

5.6.5 Image transformations

Transformations of Landsat image data are used in order to combine the information present in the four MSS or seven TM bands in some particular way. Recall that the spectral reflectance curve of vigorous, healthy vegetation shows a dip in the red region of the visible spectrum and a peak in the short-wave infrared (Figure 5.8(a)) whereas the corresponding curve for water shows a decline from visible green through visible red to a low in the short-wave infrared. Clouds and fresh snow have high values in all bands. A straightforward

Figure 5.19   Classified image of part of Norfolk, eastern England. The darkest areas are coniferous forest. The brightest areas are those of intensive cultivation, and intermediate shades represent grassland and sugar beet. Normally a classified image would be colour-coded as the eye is not capable of distinguishing more than six or seven levels of grey.

transformation, combining information in MSS band 7 (short-wave infrared) and MSS band 5 (visible red) is the *ratio* of the band 7 to the band 5 value, computed for each pixel. A high ratio value implies that the band 7 value of the pixel is large relative to the band 5 value while a value of 1 would indicate that the band 5 and band 7 values were equal. The ratio technique has been used to assess the vigour of growing vegetation not only using Landsat data for small area inventories but also on a global basis using low-resolution data from the NOAA meteorological satellites. The AVHRR sensor carried by the

NOAA satellites has a visible and a short-wave infrared band, and the National Oceanographic and Atmospheric Administration in Washington, DC produces global vegetation maps on a routine basis using the ratio transform outlined above.

A second widely-used transform is that of *principal components*. Principal components are defined in terms of combinations of the pixel values in each of the four MSS or seven TM bands. These combinations are computed in such a way as to concentrate the maximum amount of information in principal component 1 and the least in principal component 4 (or 7 if TM data are used, for there are as many principal components as there are bands of imagery). The combinations are of the form

$$C = (a \times MSS4) + (b \times MSS5) + (c \times MSS6) + (d \times MSS7)$$

where C is the value of the principal component for a particular pixel, and a, b, c and d are coefficients derived from the correlations between the bands. The terms MSS4–MSS7 are the values recorded in MSS bands 4 to 7 for the pixel concerned.

Where the spectral bands making up the multispectral TM or MSS image are highly correlated then more of the information contained in the multispectral image will be compressed into principal component 1 than would be the case if the intercorrelations among the bands were low. The first principal component for a set of four MSS bands generally accounts for 85–90% of the total information in all the bands. It is usually thought to represent "average reflectance" while principal component 2 (accounting for 5–20% of the total information in all four bands) is normally associated with the visible/near-infrared contrast. This contrast between high infrared and low red values is indicative of vigorous vegetation growth, hence principal component 2 is often labelled "greenness". Components 3 and 4 are much less important and are dominated by noise, though this is not to say that the small amount of information they contain is irrelevant.

Figures 5.20(a–d) show the four principal components derived from analysis of a Landsat MSS image of a semi-arid area in Northern Chile. The variation in reflectance is largely due to differences in rock type and intensity of surface weathering. The decline in information content from principal component 1 to principal component 4 does not need any verbal description. Component 4 is dominated by the scan line banding mentioned previously, but some anomalous dark areas may be worthy of investigation. Component 1, on the other hand, is free from obvious banding and it is possible to pick out several areas that differ in surface reflectance and degree of dissection. This image would be useful in preparing a reconnaissance geological survey of this inaccessible and inhospitable region. The technique is, in fact, widely used by geologists who often combine principal components 1 through 3 to form a principal components colour composite image for the study of rock types.

Figure 5.20 **(a)–(d)** Principal components 1–4 respectively of a Landsat Multispectral Scanner (MSS) image of part of Northern Chile. The first principal component has most information and hence the greatest contrast and least noise. The pattern of light and dark areas represents variation in rock type and surface weathering in this arid region. The dark area in the top right of Figure 5.20(a) is seen more clearly as a bright region on Figure 5.20(b). It is a small lake. The noise level increases considerably in Figures 5.20(c) and (d), showing that most of the information in the four bands of the MSS image can adequately be represented by two principal components.

5.7 SUMMARY

The techniques of display, enhancement, filtering and transformation of remotely-sensed images that have been described in this chapter represent only a selection of those available. As experience of Landsat image processing

Figure 5.20 (*cont.*) (**b**)

accumulates so geographers will become more adept at extracting information from such images. Remotely-sensed images will become an increasingly important source of spatial information, to be used either alone or in combination with other spatial data in the context of a Geographical Information System, as the number and variety of remote sensing satellites is increased. France, Japan, Canada, India, the USSR and Europe all have active remote sensing programmes, and US dominance of the commercial remote sensing market received its first challenge in 1986 with the launch of the French Satellite Pour l'Observation de la Terre (SPOT-1). This satellite has a spatial resolution three times that of Landsat's TM and also has the ability to generate oblique, rather than vertical, views. Stereoscopic images can be produced from such oblique

Figure 5.20 (*cont.*)   (**c**)

views, thus improving the information content of the images by allowing terrain
elevations to be derived directly from the remotely-sensed images. SPOT-2 was
successfully launched in 1990. A Japanese remote sensing satellite, designed for
marine observations, was launched in February, 1987. Other satellites, such as
the European ERS-1 and the Canadian Radarsat, are scheduled for the period
1990–1995 and will provide radar imagery of the Earth for the monitoring of
storms at sea and the tracking of iceberg movement in Arctic waters. The volume
and quality of remotely-sensed data seem set to increase substantially in the
years to come, and the manipulation and interpretation of remotely-sensed data
will, correspondingly, become a valuable geographical skill.

Further information on the material covered in this chapter can be found

Figure 5.20 (*cont.*) **(d)**

in Curran (1985) and Harris (1987) at an introductory level. Mather (1989) gives more details of computer techniques described in this chapter.

5.8 REVIEW QUESTIONS

1. Define the following terms:

Landsat	ground control point	contrast stretch
Multispectral Scanner	waveband	registration
scattering	infrared	pixel
lowpass filter	atmospheric window	

2. What are the advantages to the geographer of synoptic and repetitive imagery of the Earth's surface from orbital altitudes?

3. Outline the difficulties that might be encountered when relating Landsat imagery to conventional maps. What are the main factors producing geometric distortion in Landsat images?

4. What is meant by the term noise? What are its effects on the visual interpretation of remotely-sensed images? What methods can be used to eliminate or reduce noise?

5. Define the term *thematic map*, giving examples. What computer processing methods are used to derive thematic maps from multispectral remotely-sensed images? What phenomena can be mapped from such imagery?

6. Use a simple numerical example to show how the ratio between Landsat MSS band 7 (short-wave infrared) and band 5 (red) can distinguish between areas of vigorous vegetation and unvegetated areas. What practical difficulties might you encounter in deriving a digital ratio image? (*Hint*: a computer, like a human being, cannot divide by zero. Also, a memory bank forming part of an image display system holds whole numbers, not fractions.)

CHAPTER 6

Simulation

6.1 INTRODUCTION

Teaching a trainee pilot to land an aircraft with a failed engine is a hazardous activity. It is not surprising that airlines don't train their pilots on real aircraft, otherwise they might quickly run out of both pilots and aircraft (as well as instructors). Consequently, most pilots are trained on flight simulators. The trainee feels that he or she is flying a real aircraft, and can see the view from the cockpit just as if (s)he were actually approaching the runway at a particular airport. However, if a "crash" occurs the worst that can be expected is a rebuke from the instructor. The term *simulate* is used in this chapter to mean mimic or copy a real event or situation. The two reasons for using simulation methods in pilot training are to reduce costs and to eliminate hazards.

Simulation of geographical systems is not necessarily motivated by considerations of cost or avoidance of hazard, but more by the desire to experiment on systems that are either too slow-acting relative to the human life-span or which present ethical problems to the would-be experimenter. In this context a system is defined to be a set of inter-related components together with the relationships between them (Huggett, 1980). An example of a slow-acting system is a hill-slope. In terms of a human life-span, the development of a hill-slope takes a very long time, and one could not stay around long enough to test alternative theories of hill-slope development if observation of processes acting on the present landscape produced the only relevant data. A system that presents ethical problems to the experimenter is the demographic system. Without the absolute power of an Egyptian Pharaoh one could scarcely test the consequences of an experiment involving the manipulation of people's lives, for example forcing a percentage of the population to emigrate or insisting that each fertile married couple produce exactly two offspring. Other examples of applications of simulation methods in geography include the prediction of the hydrological behaviour of a river or the travel patterns of consumers in a region where a new shopping centre is to be built.

177

An introduction to the use of computer simulation models in geography is presented in this chapter. Study of the examples will show that in order to simulate the behaviour of a system one must in the first instance understand it. The design and testing of computer simulation models provides a valuable way of testing the consequences of an idea or theory. The first example uses a simulation model of a simple pattern, that of rivers in a drainage basin.

6.2   RIVER PATTERN SIMULATION

Imagine a drunken man attempting to walk across a large, empty car park. If matters are simplified for him by assuming that he can move only north, south, east or west, and excluding the possibility that he might collapse on the spot, can you visualize the path he would follow? It would be totally random, so it could be called a *random walk*. Surprising as it might seem, the random walk model is used in science to simulate situations in which there are no dominant, controlling forces. This implies that the state of the system at successive time-periods, such as the successive positions of the drunken man, are uncorrelated. If a very simple model is taken, representing stream patterns developing on a uniform surface with little or no slope, then no dominant process would control the direction of stream flow. The random walk model could then be used to build up a picture of the possible range of river patterns that might develop in such circumstances. This model is unrealistic, but is used here to demonstrate the principles involved.

The simplest case of a single stream is considered first. The area of interest is divided into a set of rectangular grid cells; each cell is identified by a pair of row/column coordinates so that, for example, cell (5,2) is the second cell along the fifth row from the bottom of the grid. The first assumption is that each cell can accommodate only a single reach or segment of a stream, and the stream can therefore move through each cell only once. The second assumption is that the stream cannot move backwards on itself, that is, reverse its direction or join itself. In view of the observed behaviour of natural streams, these assumptions are realistic. The third assumption is that the stream terminates at the edge of the grid.

The simulation begins by the selection of the coordinates of the cell in which the stream has its source, and proceeds by random selection of the direction of flow. These operations could be performed manually by the use of a dice. If the maximum number of cells in a row of the grid is 30 and if there are 30 rows, a dice could be thrown six times and the outcomes totalled to get the row coordinate, and six times for the column coordinate using the convention that the six on the dice is interpreted as zero. In the unlikely event that six successive sixes (zeros) were thrown the result would be counted not as zero but as one, or the dice might be thrown again. Next, the dice could be used to determine the direction of flow from the source cell. The outcomes 1, 2, 3 and 4 could be interpreted as north, south, east and west with outcomes 5 and 6 being ignored. The dice would be recast if the outcome was incompatible with the assumptions, so that a southerly-flowing segment would be prohibited from turning through 180° and flowing north. The game ends when the stream reaches the boundary of the grid. Figure 6.1 shows a possible outcome for an 11 × 14 grid.

No single result of a simulation experiment can be considered meaningful. Just as a statistical sample (Chapter 5) must be of a certain size before reliable inferences concerning the population can be drawn, so a number of simulations must be run before the average behaviour can be treated as being descriptive of the real system. However, carrying out a number of simulations by hand is both time-consuming and tedious. On the other hand, if the simulation is to be carried out by computer a way must be found of getting the computer to generate random sequences of numbers such as would be obtained by throwing a dice a number of times. A computer is a machine with the attractive property that it does exactly as it is told, yet the outcome of a series of throws of an unbiased dice will produce a random sequence of numbers in which the $n + 1$th outcome is independent of the nth. How can a series of fixed instructions cause the computer to come up with a random sequence of results? The answer is, of course, that it cannot. However, computer scientists have developed programs which produce numbers that are apparently random. They are called *pseudo-random* numbers and most computers have built-in functions which will generate sequences of such numbers. In the BASIC language this function is called RND. The possibility of using the computer to generate apparently random sequences of numbers allows the production of computer programs which, in essence, copy the human ability to throw an unbiased dice or toss a coin.

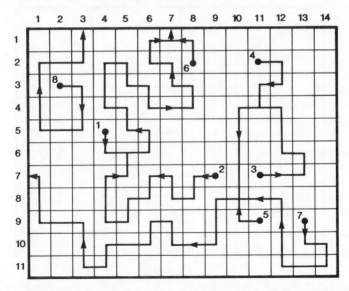

Figure 6.1   Randomly-generated stream pattern. The stream sources (numbered 1 to 8) are chosen randomly, as is the direction of stream flow.

A computer program to simulate the simple stream-pattern experiment described earlier would follow the same lines as a manual simulation. The source coordinates of the stream are found by generating two random whole numbers in the range 1–30 and the direction of movement from one cell to the next is found by generating another random number which can take on one of the values 1, 2, 3 or 4. It is relatively easy to prevent computer-generated streams from altering direction through 180° (the first assumption) simply by comparing the present direction of flow (1, 2, 3 or 4) with the newly-generated random number. The newly-generated number cannot be a 2 (south) if the current flow direction is 1 (north), nor can it be a 3 (east) if the current direction of flow is 4 (west), and so on. It is more difficult to prevent the river going round in a loop and joining itself. If, though, we place a zero in each cell of the grid before the simulation starts we can use the code '0' to mean that the cell is unoccupied. The presence of a stream segment in a grid cell is indicated by the code '1'. The program could then incorporate a rule which prevented a move into a cell containing a non-zero code. Sometimes the stream will develop to a position in which it cannot move anywhere without violating one of the two rules (no flow reversal and no joining itself). The program would then run forever. To escape from such a condition it must first be recognized and then a strategy to move backwards a step at a time until a new route is found must be followed. For example, in the situation shown at cell (5,7) in Figure 6.2 the stream is moved

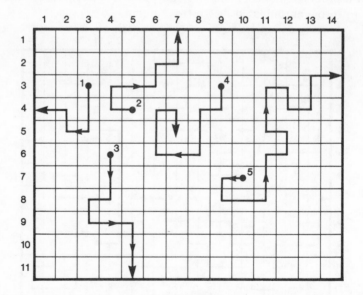

Figure 6.2   The stream pattern shown in Figure 6.1 is subject to two constraints: a stream cannot flow backwards, nor can it join itself. Stream 4 in this diagram violates the second constraint at cell (5,7).

back one cell towards the source, without altering any of the 1s and 0s, until the stream can move in a different direction.

A BASIC program running on a microcomputer can generate a single stream pattern in about one second. It is relatively easy to extend the program so that several streams are generated. The only modifications to the logic discussed above are that the code for "square occupied" is no longer 1, but is the actual stream number (1 for the first stream, 2 for the second and so on). A stream ends either when it reaches the edge of the grid or when it joins another stream.

The multiple-stream model allows the simulations of real stream patterns, using measurements of actual streams as a guide. The first hypothesis is that, in the absence of any controls (such as slope, lithology or structure), a stream pattern will develop randomly and will be described by the random-walk model. To test this hypothesis a number of stream patterns are generated using the program described above and a count made of the number of streams with no tributaries (these are called first-order streams), then the number of streams formed by the junction of two first-order streams (these are second-order streams) and so on. Figure 6.3 shows how streams are ordered. The geologist A. N. Strahler showed that if the logarithm of the number of streams of each order is plotted against stream order then a straight line will fit the resulting scatter of points. If a similar plot for a set of randomly-generated stream patterns shows the same result then it can be concluded, that in the absence of geological and topographical controls, stream patterns appear to be generated by random

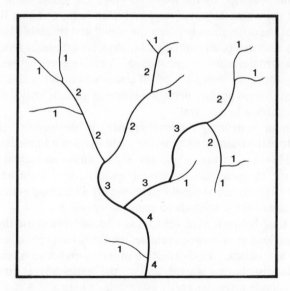

Figure 6.3   Strahler's method of stream ordering.

processes. A random process is often considered to be the sum of a large number of small, independent processes so this reasoning is acceptable.

Further hypotheses might be considered; for example, the effect of slope could be incorporated into the model by assigning different probabilities to each of the directions of flow. A southerly slope could be simulated by ensuring that 50% of the random numbers generated by the computer were interpreted as "flow south" while 20% could be allocated both to "flow east" and "flow west" with the remaining 10% being "flow north". Practical tests on this type of simulation model have shown that real stream patterns are similar (in terms of the characteristics incorporated in the model) to those developed by simulation techniques. The use of simulation allows the observation of many possible stream patterns, far more than could be observed in the real world.

6.3   LEVEL CROSSING SIMULATION

This section is partly based on an example from Guttman (1977, p. 174). It concerns the development of a simulation model of a railway level crossing which, in the simplified version presented here, assumes a single railway line crossing a road on which traffic moves in one direction only. An extended version of the model might be used to simulate the effects of the crossing for different road and rail traffic conditions. In the present version of the model the train timetable is fixed for a 24-hour period, with trains running every x minutes, x being a value specified by the user. To make the model more realistic it is assumed that the trains do not always arrive on time, and that the frequency distribution of the differences between the actual and timetabled arrivals of the trains follows a normal (Gaussian) distribution with a mean of zero (Chapter 3). If s is the standard deviation, in seconds, of the differences between actual and scheduled arrival times then 68% of all trains will arrive within s seconds of their scheduled time. A positive deviation indicates an early arrival and a negative deviation indicates a late arrival.

The signalman controlling the level crossing is instructed to close the gate n seconds before the train actually arrives, and to open it immediately the train has passed. However, if the next train is due within m seconds of the gate reopening time, the gate remains closed until the next train has gone. It is assumed that all trains are of equal length and are all moving at the same speed, so that they each take p seconds to pass the crossing.

The mean time between road vehicles is r seconds. Because there are more short gaps than long gaps between vehicles the Gaussian probability distribution is not used to approximate the distribution of time gaps between vehicles. Instead an exponential distribution is used, so that the actual arrival time of the next vehicle is x seconds after the previous vehicle, where x is a random variable sampled from an exponential distribution with a mean of r seconds. Short gaps

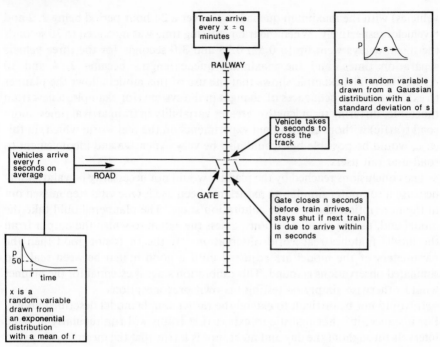

Figure 6.4 Level-crossing simulation model in graphic form. The purpose of the simulation is to estimate mean vehicle queue lengths and delays for various vehicle densities and train frequencies.

are thus more probable than long ones. Each vehicle takes b seconds to cross the railway tracks. The model is shown graphically in Figure 6.4

Variables such as x and b in the preceding description are all specified by the user of the program, so that different situations can be modelled. In this example the following specification is used: trains run regularly at 15-minute intervals throughout the 24-hour period. The standard deviation of the differences between actual and scheduled train arrival times is 90 seconds. The gate closes 30 seconds before the arrival of a train, and remains closed if the next train is due within a further 90 seconds. Each vehicle takes 5 seconds to travel over the crossing. The mean time between vehicles and the time taken by the train to cross the road are the factors that are varied; the following mean times between vehicles were used: 70, 35 and 10 seconds. For each of these values there were two train crossing times: 10 and 30 seconds. Six simulation runs were thus possible, with two train crossing times for each of the three vehicle separation times. The results showed that, for a train crossing time of 10 seconds, the mean delay to road vehicles was 0.14 second (70 seconds between vehicles), 0.23 second (35 seconds between vehicles) and 1.8 seconds (10 seconds between vehicles) with the maximum queue length over a 24-hour period being 2, 2 and 8 vehicles respectively. When the train crossing time was increased to 30 seconds the mean delays went up to 0.52, 0.75 and 3.0 seconds for the three vehicle separation times, and the maximum queue lengths became 2, 4 and 10 respectively. The example shows that the use of this model allows the planner to examine the consequences of changes in the system (for example, a new train timetable, differing train lengths, greater variability in train arrival times, more road traffic) without carrying out experiments on the real world which, in this case, would be possible but would also be very expensive and inconvenient to road and rail users.

The conclusions reached by the planner would not necessarily be valid simply because a computer simulation model had been used. One vital step missed out in the description above is the calibration stage. The planner should take the model and, using real observations, assess the extent to which the output from the model matches real-world observations. If the fit is not good then the parameters of the model are adjusted until a good match between real and simulated observations is found. This calibration stage is essential for the planner would otherwise simply be testing his own preconceptions.

It would not be difficult to extend the rather simple model described above. For instance, it is not realistic to expect that trains will run regularly at fixed intervals throughout the day and night, nor is it true that the mean time between vehicles is the same over a 24-hour period. Additional programming would be needed to permit (a) the real train timetable to be entered and used by the program and (b) changes in the mean time between vehicles to take place at hourly or half-hourly intervals throughout the day to allow a closer representation of reality. However, the simple model outlined above shows the stages of

- specification of the properties of and relationships between the components of the model,
- calibration of the model, and
- validation of the inputs to the model in order to assess the effects of such changes on the behaviour of the model.

6.4   MARKOV CHAIN MODELS

The random-walk model described in Section 6.2 is based on the assumption that each possible state of the system (the stream flow direction in the example in Section 6.2) is equally probable. In some geographical systems this is not the case, and the next state of the system is related to its present state. This feature is called the *Markov property* and a sequence of observations possessing this property is called a *Markov chain*. The state of the weather on a particular day could be taken as an example. The weather might be described as "wet" or "dry" (W or D) and, in a humid temperate climate such as Britain experiences, it might seem that state W is more common than state D. The reverse may apply in a semi-arid area such as Arizona or Nevada. The occurrence of wet and dry days is thus not random – there is a persistence effect such that a wet day is more likely to be followed by another wet day than by a dry day, and vice-versa. This is due to the duration and movement patterns of the weather systems (depressions and anticyclones).

If the state of the weather over a long sequence of days were observed, and record W or D for each, one could work out how often

- a W is followed by another W,
- a W is followed by a D,
- a D is followed by a W, and
- a D is followed by another D.

These four counts are transformed into estimates of probability by dividing each by the total number of observations. If there were 1000 observations altogether, and 500 represented state 1 in the list above (W followed by W) then the probability of a wet day following a wet day would be estimated as 0.5. The four estimated probability values are conventionally represented as shown in Table 6.1. In this table, p(1) is the probability that if it is raining today (time t) it will be raining tomorrow (time $t + 1$), while p(2) is the probability that if it is raining today it will be dry tomorrow. Probabilities p(3) and p(4) estimate, respectively, the likelihood of (i) dry today, wet tomorrow and (ii) dry both today and tomorrow.

Another example of a Markov chain is the movement pattern of a business executive who visits four cities each week. The order in which the visits occur is determined by the executive. The travel pattern might be influenced by such factors as convenience, minimizing travel time and the nature of the business (for example, the businessman might wish to visit suppliers, manufacturers and retail outlets, in that order). If the four cities are labelled a, b, c and d then, in order to estimate the probability that the businessman will travel from a to b on successive days (or c to a, or any other combination, including staying consecutive days in the same place) then some observations of his past travel pattern are needed. These observations are converted to probabilities just as the weather data were converted to probabilities in the first example. The table

Table 6.1   Transition probability matrix for wet (W) and dry (D) days

Today	Tomorrow	
	W	D
W	p(1)	p(2)
D	p(3)	p(4)

Table 6.2   Transition probabilities for the business executive's
travel pattern

	Next town visited			
Current location	a	b	c	d
a	p(1)	p(2)	p(3)	p(4)
b	p(5)	p(6)	p(7)	p(8)
c	p(9)	p(10)	p(11)	p(12)
d	p(13)	p(14)	p(15)	p(16)

of transition probabilities for the travel example would be laid out as shown
in Table 6.2.

Transition probabilities p(1) to p(4) in Table 6.2 are calculated by counting
the number of times the business executive stayed in city a and subsequently
(i) stayed the next day in city a, (ii) moved to city b, (iii) moved to city c,
(iv) moved to city d. If the counts were 7, 2, 3 and 9 then we first find the sum
of the counts (21) and divide each individual count by the sum of the counts
to give p(1) = 0.33, p(2) = 0.09, p(3) = 0.14 and p(4) = 0.43. The sum of these
probabilities should be 1.0, meaning that the executive must do one of the four
things mentioned. In fact the sum is 0.99, the error being due to rounding to
two decimal places. From the four calculated probabilities it can be deduced
that if the executive is currently in city a then the chance that he or she will
(i) stay in city a is 33%, (ii) move to city b is 9%, (iii) move to city c is 14% or
(iv) move to city d is 43%. Probabilities p(5) to p(16) are found in a similar way.

Matrices of transition probabilities such as those given in the earlier examples
are useful in that they summarize the behaviour over time of the system under
study. Transitions or changes which have a notably high or low probability can
be identified, as reasons for their prominence sought. Secondly, through the
use of techniques of Markov chain analysis, the equilibrium state of the system
can be established. The equilibrium state describes the proportion of time the
system spends in each state in the long term. Thirdly, the table of transition
probabilities can be used as the basis for simulation, for many different
realizations of a process (travelling between four cities, or daily rainfall patterns)
can lead to the same set of transition probabilities. Each of these realizations
is equally likely. Such simulations, based on observed system behaviour, can
tell us how things might have been if we are looking at unrepeatable, past, events
or how the system might behave in the future, if the transition probabilities
remain unaltered. Thus, we might look at lithological changes through the
geological column. Changes in the state of the system, in this case changes from
one rock type (for example, shale) to another (for example, grit) could be
examined and counted in bore-holes and outcrops. The Markov chain model
might then be used to draw inferences about the lithological system from these
sample observations.

Table 6.3   Structure of the table of counts
(transition matrix) for the occurrence of clear
and foggy days in Los Angeles

	Today	
Yesterday	Foggy	Clear
Foggy	n1	n2
Clear	n3	n4

To illustrate the points made above some data obtained from a study of the occurrence of clear and foggy days in the area of Los Angeles, California, carried out by Gong-Yuh Lin (1981), are used as an example. From the data obtained at a number of air-monitoring stations, Lin produced a table of counts in the form shown in Table 6.3. When the actual counts were converted to transition probabilities the results shown in Table 6.4 were obtained.

Table 6.4 shows that on about 75% of occasions the state of the atmosphere (foggy or clear) remained the same from one day to the next, and a change from one state to another occurred on about 25% of occasions. The table can be used to estimate the average proportion of days which, in the long term and if these probabilities are representative, will be foggy or clear. These proportions are found by the technique of powering the table of probabilities. By powering is meant raising to the power of; thus, the single (scalar) quantity two can be raised to the power of three to give the answer eight. A matrix is not a simple scalar quantity like 2 or 8 and a special technique is needed to carry out the operation of raising a matrix to the power n. First of all, note that (in scalar arithmetic) we can find x to the power of y by a series of steps: first, multiply x by x to give x^2. Next, multiply x by x^2 to give x^3, and so on until x has been raised to the power y (x^y). The probability matrix is raised to successively higher powers in much the same way. To raise a matrix to the power of two, the layout shown in Table 6.5 is used.

Table 6.4   Transition probabilities for
Californian foggy/clear day example

	Today	
Yesterday	Foggy	Clear
Foggy	0.77	0.23
Clear	0.24	0.76

Table 6.5   Layout   used   for   matrix
multiplication

$$\begin{bmatrix} p_1 & p_2 \\ p_3 & p_4 \end{bmatrix} \times \begin{bmatrix} q_1 & q_2 \\ q_3 & q_4 \end{bmatrix}$$
$$= \begin{bmatrix} r_1 & r_2 \\ r_3 & r_4 \end{bmatrix}$$

The matrix $\mathbf{P}$ with elements p_i is multiplied by the matrix $\mathbf{Q}$ with elements q_i to give the results matrix $\mathbf{R}$ with elements r_i. Each element of the results matrix is obtained by multiplying the elements of a row of matrix $\mathbf{P}$ by the corresponding column elements of matrix $\mathbf{Q}$ and adding the results. In this case,

$$r_1 = p_1{}^*q_1 + p_2{}^*q_3$$
$$r_2 = p_1{}^*q_2 + p_2{}^*q_4$$
$$r_3 = p_3{}^*q_1 + p_4{}^*q_3$$
$$r_4 = p_3{}^*r_2 + p_4{}^*q_4$$

If this procedure is applied to the transition probability matrix given earlier then the result shown in Table 6.6 is obtained.

The elements of $\mathbf{R}$ (r_1, r_2, r_3, r_4 are calculated as explained earlier; the result is $r_1 = 0.65$, $r_2 = 0.35$, $r_3 = 0.37$ and $r_4 = 0.63$. Matrix $\mathbf{R}$ is the transition probability matrix raised to the power two. In order to raise the transition probability matrix to the power three the layout of Table 6.7 is used to get $\mathbf{S}$. $\mathbf{S}$ is equivalent to $\mathbf{P}^3$, for $\mathbf{R}$ ($=\mathbf{P}^2$) is multiplied by $\mathbf{P}$ to give

$$\mathbf{S} = \mathbf{P}^3 = \begin{bmatrix} 0.58 & 0.42 \\ 0.44 & 0.56 \end{bmatrix}$$

The values in the first column of $\mathbf{P}^3$ (0.58 and 0.44) are closer together than the corresponding values in the matrix $\mathbf{P}^2$ (0.65 and 0.37) and much closer than the values in column one of matrix $\mathbf{P}$ (0.77 and 0.24). The same is true of the second column – the two elements making up that column are getting closer together in value as $\mathbf{P}$ is raised to successively higher powers. Eventually the two values in column one become effectively equal as do the two values in column two. In this example, convergence occurs when the two values in column one

Table 6.6   Showing the method of finding the matrix $\mathbf{R} = \mathbf{P}^2$

$$\mathbf{P} = \begin{bmatrix} 0.77 & 0.23 \\ 0.24 & 0.76 \end{bmatrix} \times \mathbf{P} = \begin{bmatrix} 0.77 & 0.23 \\ 0.24 & 0.76 \end{bmatrix}$$

$$= \mathbf{R} = \begin{bmatrix} r_1 & r_2 \\ r_3 & r_4 \end{bmatrix}$$

Table 6.7   Layout to find the matrix $\mathbf{S}$ ($=\mathbf{P}^3$)

$$\mathbf{R} = \begin{bmatrix} 0.65 & 0.36 \\ 0.37 & 0.63 \end{bmatrix} \times \mathbf{R} = \begin{bmatrix} 0.77 & 0.23 \\ 0.24 & 0.76 \end{bmatrix}$$

$$= \mathbf{S} = \begin{bmatrix} r_1 & r_2 \\ r_3 & r_4 \end{bmatrix}$$

both become equal to 0.51 and the two values in column two become equal to 0.49. This result is interpreted to mean that, in the long run, the system (atmospheric state) will be in state one (foggy) 51% of the time and in state two (clear) 49% of the time. Because the calculations involved in powering matrices are repetitive they are ideal for the computer and so Markov chain analysis is normally performed with the aid of a computer program.

Many different patterns of foggy and clear days could produce the same numerical values for the transition probabilities in the example. It is sometimes interesting to use the transition probability matrix to generate synthetic sequences of foggy and clear days; these sequences show patterns that might be expected to occur in the future (since there are an infinite number of alternative futures but only one present and one past). Study of these alternative future sequences might show a particular tendency, with one type of pattern being more prevalent (and hence more likely) than others. The steps involved in generating a synthetic sequence are:

1. Decide arbitrarily whether the initial state of the system is F (foggy) or C (clear).
2. Generate a pseudo-random number pr (Section 6.2) in the range 0.0–1.0. If the present state is F then if pr is less than 0.77 then the state remains F or else the state becomes C. If the present state is C then if pr is less than 0.24 then the state becomes F or else the state remains C. The values 0.77 and 0.24 are obtained from the transition probability matrix shown in Table 6.4.
3. Repeat step 2 until the desired number of realizations of the system has been attained.

Here are four synthetic sequences generated from the foggy/clear days transition probability matrix given above:

```
FFCCCCFCFFFFFFFFFFCFCCFFFFFFFFFFF
FFCCFFFFFCFFFFFFFFCCCCCCCCFCFFCC
CCFFCFCCCCCCCCCCCFFFFFCFCFCFFFF
CCCCCCFFFCCFCFFFFFFFFFCCCFFFCCFF
```

It is apparent from these sequences that the persistence of a particular state (F or C) is considerable; if one day is foggy then the next day is more likely to be foggy than clear, and vice-versa. This is what the transition probability matrix tells us. The length of the synthetic sequence should be long enough for a pattern to emerge; the length of the four sequences given above is not sufficient for reliable conclusions to be drawn. For example, study of the third sequence might lead us to believe that foggy days are relatively rare occurrences at Upland, whereas the powered transition probability matrix gave a result of 51% foggy days and 49% clear days.

The example shows how a system which can be in one of two states can be simulated, and how the results of these simulations can be used to gain some understanding of system behaviour and of the controls which influence this behaviour. Real systems are rarely so simple; their behaviour can normally be described only in terms of many variables. In the final sections of this chapter, examples of more complex systems are used, and consideration is given to the role of computer simulation techniques in understanding and predicting their behaviour.

6.5   POPULATION FORECASTING*

The advent of the computer has made possible the development of complex simulations of situations in human and regional geography. The great speed with which the computer carries out calculations means that experiments can be carried out quickly and cheaply. The availability of on-line facilities and displays allows the inspection of results as a simulation proceeds through time and gives the operator the opportunity to make decisions about progress at regular intervals. Perhaps the most important consideration in making any simulation that uses a computer program is to know exactly what steps and calculations are needed or, in other words, how a particular system works. It is therefore necessary to know how the component variables of the system interact with each other. Often the construction of a simple flowchart or systems diagram can help in this phase of comprehension. In order to illustrate the procedure and problems of designing a simulation an example of changes in a human population will be taken.

The simplest way in which these changes may be projected through time is to take the total population at a given instant and to alter it year by year (or by some other period of time) according to a chosen assumption about future growth or decline. For example, in a particular year a given region has x million inhabitants and the population is expected to increase at a rate of 2% per year. To anticipate future populations, given such a rate of increase, a compound interest-type calculation is an obvious possibility.

It is often useful in the study of population to distinguish between males and females and also between different age-groups. If forecasts of future population at 10-year intervals are needed, then the population can be divided into nine age-groups each representing a 10-year age-group or cohort (0–9 years, 10–19 years, and so on, to 80 plus). More refined subdivisions of population by age are by five-year or even one-year age-groups.

The simulation proceeds by computing, for each age-group and for males and females separately, the number of deaths. In any real-world population the death rate changes from one age-group to another while, for a given age-group, the male and female death rates are also different. The number of deaths

---

*This section was contributed by Prof. J. P. Cole

computed for each age-group is subtracted from the population in that age-group before the number of births is computed. The number of births depends on the number of females of child-bearing age, and the ratio of male to female births may be (and is normally) rather greater than 1 : 1. The number of births for males and females separately is added to the population of age-group 1 (0–9 years in the example above). When an age-group limit is reached (every 10 years in the case of 10-year age-groups, or every five years in the case of five-year age-groups) then the population structure is updated by moving every age-group forward by one group. The members of the last age-group are assumed to have a mortality rate of 100%. A flow-chart of the model described in this paragraph is shown in Figure 6.5.

The results of a simulation based upon the model described above are shown in Figure 6.6. Figure 6.6(a) shows the population structure of Peru in 1983 in the form of a population pyramid in which each of the rows represents a 10-year age-group with the lowest row representing the youngest age-group. The symbols M and F stand for males and females. If the simulation is run using estimated early-1980s fertility and mortality rates the population in 2003 would be as shown in Figure 6.6(b). If, however, fertility rates fall for the four child-bearing age-groups (10–19, 20–29, 30–39 and 40–49) from the estimated early 1980s figures of 0.4, 2, 2 and 0.4 children per woman respectively to 0.25, 1.8, 1.4 and 0.3 children per woman then the population structure in 2003 would be shown in Figure 6.6(c). The effects of different fertility and mortality rates can be assessed quickly, and our understanding of population growth can thereby be enhanced.

The model described above is relatively simple. It does, however, represent a basic structure which can be extended and refined by, for example, subdividing the population into urban and rural components or considering the interactions between populations of several regions. The effects of migration can also be allowed for, with appropriate assumptions made about numbers or proportions of population in the different age-groups and the two sexes moving between regions. By now our model is beginning to get fairly complex. A danger of including too much in a simulation is that the results get very extensive and perhaps difficult to comprehend.

The basic population simulation outlined above is adequate as a starting-point. When applied to forecasts of present population it can reveal interesting possible futures or help to confirm more precisely suppositions already made. If the recent pattern of population change observed in many developing countries were to continue without modification then, in a few decades, there would be some impossibly large populations. On the other hand, without the use of a simulation model it is difficult to visualize the effect that changes in fertility or in mortality are likely to have on trends over the next few decades. It is possible also to study the effects of drastic changes in existing fertility and mortality rates. For example, planners in China do not seem aware of the cohort structure their population might have in decades to come if the drastic reductions in fertility

192

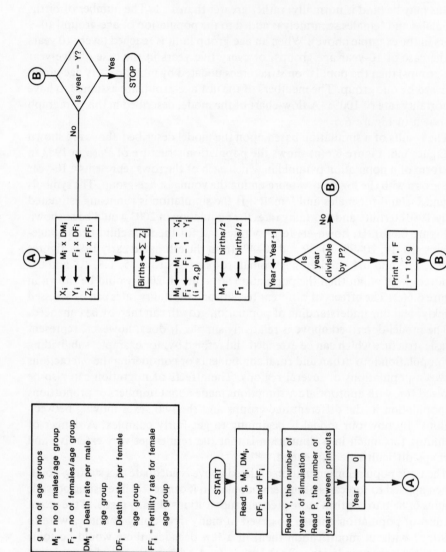

Figure 6.5   Flow-chart of the population projection model.

```
                              !
                              !F
                          MM!FF
                        MMM!FFF
                      MMMMM!FFFFF
                    MMMMMMM!FFFFFFF
                  MMMMMMMMMM!FFFFFFFFFF
              MMMMMMMMMMMMMMMM!FFFFFFFFFFFFFFF
          MMMMMMMMMMMMMMMMMMMM!FFFFFFFFFFFFFFFFFFFF
```
_____

                    males   females

                    _____

(a)   ! TOTAL POPULATION: 18.51m !
                    _____

```
                              !
                          MM!FF
                        MMMM!FFFF
                      MMMMMM!FFFFFF
                  MMMMMMMMMM!FFFFFFFFFF
                MMMMMMMMMMMMMM!FFFFFFFFFFFFFF
            MMMMMMMMMMMMMMMMMMMM!FFFFFFFFFFFFFFFFFFF
          MMMMMMMMMMMMMMMMMMMMMMMM!FFFFFFFFFFFFFFFFFFFFFFF
    MMMMMMMMMMMMMMMMMMMMMMMMMMMMMMMM!FFFFFFFFFFFFFFFFFFFFFFFFFFFFFF
```
_____

                    males   females

                    _____

(b)   ! TOTAL POPULATION: 38.88m !
                    _____

                    POPULATION PYRAMID FOR 2003
                    (revised ferility after 1990)


```
                              !
                          MM!FFF
                        MMMMM!FFFFF
                      MMMMMMMM!FFFFFFF
                  MMMMMMMMMMMM!FFFFFFFFFFFF
              MMMMMMMMMMMMMMMMMM!FFFFFFFFFFFFFFFFFF
          MMMMMMMMMMMMMMMMMMMMMMMM!FFFFFFFFFFFFFFFFFFFFFF
        MMMMMMMMMMMMMMMMMMMMMMMMMMMM!FFFFFFFFFFFFFFFFFFFFFFFFFF
    MMMMMMMMMMMMMMMMMMMMMMMMMMMMMMMM!FFFFFFFFFFFFFFFFFFFFFFFFFFFFFF
```
_____

                    males   females

                    _____

(c)   ! TOTAL POPULATION: 29.06m !
                    _____

Figure 6.6 **(a)** Population pyramid for Peru, 1983. **(b)** Simulated population of Peru in 2003 assuming present mortality and fertility rates. The model (Figure 6.5) assumes a closed system without migration. **(c)** Alternative population structure for Peru in 2003. The same mortality rates are used as in Figure 6.6(b) but fertility rates are reduced (see p. 191).

introduced in the 1970s are maintained for 20 or 30 years. It is also possible to get data for past populations of countries, and to make alternative projections towards the present day, thus comparing what did happen with what might have happened. In this way a greater insight into the ways in which population size and structure change may be gained.

Other possibilities include the combination of changes in population (human or animal) with other elements and influences. A situation that is basic and very widespread is the rural community in the developing world, represented by the largely self-sufficient village which is strongly affected by local environmental conditions. The following ingredients might be combined into a simulation:

- Human population, susceptible to the kinds of changes already outlined above.
- Natural resources and environment: water, soil, possibly vegetation, climate (reflected in the growing season with hot–cold, wet–dry combinations from year to year).
- Products from natural resources: crops (different crops preferring different physical conditions), pasture, livestock (as food, materials or work animals), firewood, clay for brick-making.
- Needs of the human population: water, food, firewood, clothing, shelter, plus health, educational and transport facilities as well as luxuries.
- Interactions between the community itself and places outside: in- or out-migration of population, the exchange of products, flow of information.

Such a model is of interest to geographers since it deals with man–environment relationships. Already, though, such a study of a community involving, perhaps, only a few thousand people brings in many elements, variables and causal links. How, then, was it possible to make a rather similar model for the whole world? It must be appreciated that, although the population of the world is about a million times larger than the population of a large village, the model needed to make a reasonable study of some major world trends and problems is not necessarily much larger or more complex than the village model. Indeed, while the village is an open system, influenced by many external forces, the planet Earth, as far as the human population is concerned, is a closed system apart from the fairly constant and essential energy supply from the sun.

The power of the computer was indeed appreciated and used in the late 1960s to make a model of the world system. J. W. Forrester (1971) coordinated work on such a model and gives details of the methods and findings of his work. Several basic elements were considered individually: population, non-renewable natural resources, cultivated land and agricultural capital, industrial capital, service capital and pollution. Each element itself involves several aspects. Assumed links between different elements (such as depletion of non-renewable natural resources through industrial growth, increased pollution resulting from increased industrial activity) were made in the model. One version, described by

Forrester (1971, p. 20) as "a complete diagram of the world model, interrelating the five variables – population, natural resources, capital investment, capital investment in agriculture fraction, and pollution" can be represented in a systems diagram with 43 separate elements and various links (Figure 6.7). The Forrester model gained world-wide publicity when it was described in the book *Limits to Growth* by D. H. Meadows and his co-authors (1972).

The world model described above represents a landmark in the study of global problems. Whatever the shortcomings of the assumptions made about cause and effect in the relationship between man and the environment (and man with man) the model shows how the use of a computer to carry out the vast number of calculations necessary to realize the model can allow the consideration of a very complex situation on lines not previously possible. The world model of the early 1970s set people thinking – either to criticize its inadequacies or to suggest improvements. The frightening prospects described in the *Limits to Growth* model may have driven geographers of the 1970s to turn to the problems of smaller areas.

The computer offers many possibilities to the geographer who does not aspire to the world model. Some possible areas of application are:

• the search for raw materials,
• the location of economic activity,
• the structure of transport networks,
• the minimization of transportation costs,
• selection of crop combinations, and
• war games.

While programs already exist for many of these topics it is sometimes more satisfying if one has sufficient ability in computer programming to modify existing simulations or create alternative programs.

6.6 HYDROLOGICAL SYSTEMS

Section 6.5 closed with a description of the simulation of a large, complex system, namely, the world viewed as an integrated and interrelated set of elements such as population, resources, land, capital and pollution. In this section the problem of modelling another complex, but spatially less extensive, system comprising the drainage basin of a river is considered. The river basin is a well-defined spatial unit that is of great interest to hydrologists, geomorphologists and geographers. Geographical interest in river basins stems largely from the fact that a large proportion of the world's population lives close to rivers, which act as sources of water supply for human consumption and for agricultural and industrial use. Rivers also provide a means of disposing of waste, and act as a transport medium. Rivers also flood, and such floods are sometimes very costly in terms of loss of life and destruction of property.

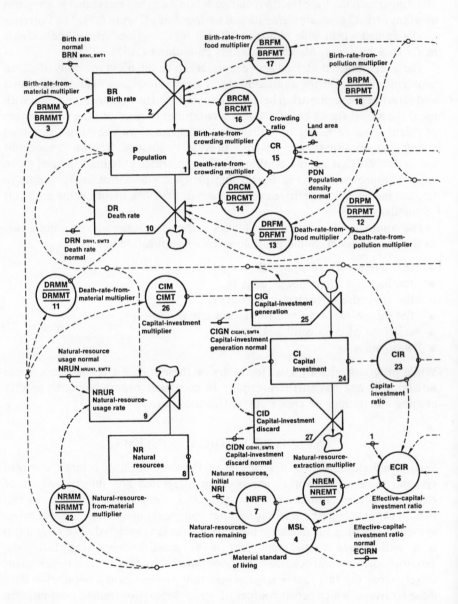

Figure 6.7 Diagram of World Model interrelating population, natural resources, capital investment, capital-invested-in-agriculture fraction and pollution. (Figure 2-1, J. W. Forrester, 1971, reproduced by permission).

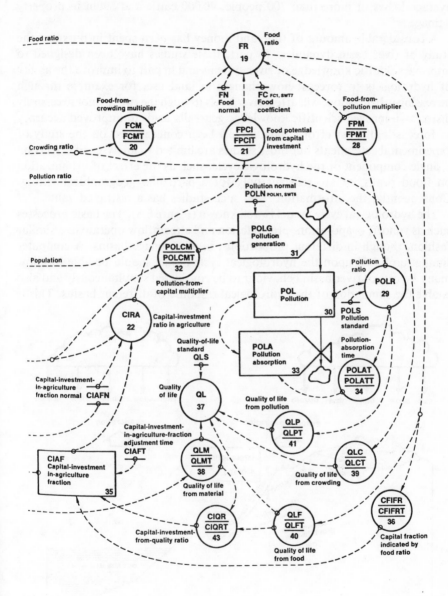

Figure 6.7 *(cont.)*

For example, the flood on the Hwang-Ho river in China in 1887 resulted in an estimated 900 000 deaths while in recent years India has suffered annual average losses of more than 700 people, 40 000 cattle and $90m in property damage.

A considerable amount of time and money has been spent in pursuing the study of river basin dynamics. In part, these studies have been designed to improve scientific knowledge of river systems and in part to improve the ability of hydrologists to forecast how changes in land use, for example through increasing urbanization, will affect river levels (though the two are not necessarily distinct – improved scientific knowledge generally leads to improved accuracy in forecasting). Past efforts have largely been concentrated on the study of experimental catchments but most studies are limited to analyses of effects on a single component of the hydrologic cycle, such as the effect of urbanization on flood peaks, or are relevant only for a particular geographical setting. Consequently, the information from such studies has a restricted value.

The hydrological cycle model is well known (Figure 6.8). The basic processes such as runoff, evaporation, precipitation and streamflow operate in a similar fashion, though in different proportions, in all drainage-basins. A computer simulation based upon the hydrological cycle model should be able to allow many aspects of river basin behaviour to be considered simultaneously and also enable the simulation of the hydrological response of different basins. This is

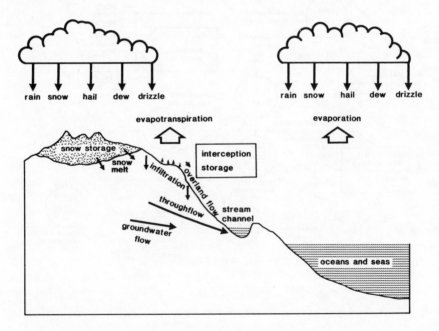

Figure 6.8   The components and processes of the hydrological cycle.

done by calibrating the model for a particular drainage-basin, that is, adjusting the model parameters until the output from the model closely matches records of the hydrological behaviour of the drainage-basin. These parameters are then used in the model to predict the behaviour of the basin.

A number of numerical models of drainage-basin hydrology have been devised. Among these are the US Department of Agriculture USDAHL model (Holtan and Lopez, 1970), the Tennessee Valley Authority Daily Streamflow Model (Betson, 1976) and the Stanford Watershed Model, which was the forerunner of the Hydrocomp Simulation Program (HSP), produced by Hydrocomp Inc., Palo Alto, California. HSP is the most powerful of the available computer simulation models. It was developed by N. H. Crawford and R. K. Linsley of Stanford University during the 1960s and has been continuously revised and updated. It can accommodate simulations of drainage-basins ranging in size from a few hectares to 40 000 square kilometres. Like other models, its use requires an understanding of the system being modelled, and so a brief description of the land phase of the hydrological cycle is considered next.

Figure 6.8 is a diagrammatic representation of the hydrological cycle. Rain, snow and hail (collectively known as precipitation) fall on the Earth's surface. A proportion is intercepted by the leaves and stems of plants, and some is evaporated back into the atmosphere. Of the water which remains, part sinks or infiltrates into the ground and the remainder flows over the ground surface. Some water collects in depressions on the surface; these depressions vary in size from the minute to lakes covering hundreds of square miles. Water flowing over the ground surface as overland flow cuts channels, and these channels combine to form a stream network. Of the water which infiltrates into the soil, some is taken up by vegetation and is transpired and the rest either flows down-slope through the soil as interflow or throughflow until it reaches a stream channel, or sinks through the soil to become part of the groundwater. Groundwater flows down-slope towards the river channel, but the rate of groundwater flow is less than the rate of throughflow or of overland flow. During dry weather, streamflow is maintained by the combination of throughflow and groundwater flow. The actual rates of operation of the processes of precipitation, interception, evaporation, transpiration, overland flow, throughflow and groundwater flow depend upon climatic, vegetative, pedological, topographic and geological conditions peculiar to each drainage basin.

Figure 6.9 shows another view of the hydrological cycle. This kind of diagram is a flow-chart or systems model. It is much more useful to the author and users of the simulation model than the pictorial representation of Figure 6.8 for it shows the inputs, outputs and storages involved in the system. The computer program must be able to take data relating to local conditions (slope, soil type and depth, amount and nature of vegetation cover, and so on) and output a streamflow record. The information about local conditions is essential if

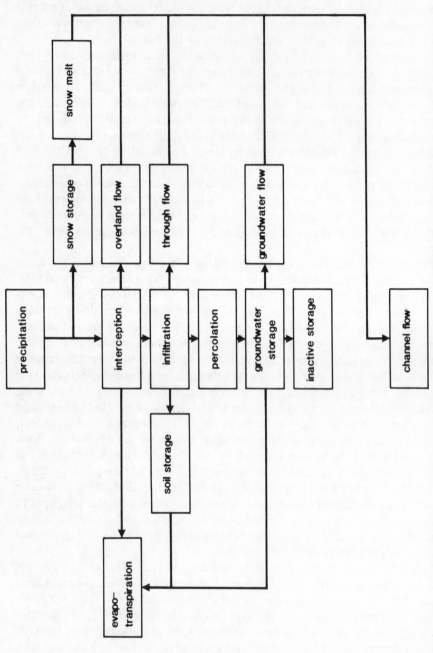

Figure 6.9   Flow diagram representing a model of the interrelationships between the processes and components of the hydrological cycle.

questions such as "if x inches of rain falls in y hours on a given drainage-basin, how will channel flow respond?" for the rates of operation of the different processes will depend on these local conditions. For example, the rates of infiltration and throughflow will depend on slope, soil type, vegetation cover, depth of soil and other factors while the volumes of water stored as depression storage will depend upon surface topography. The simulation model will also require data concerning the initial state of the system, that is, the amounts of water present in the various stores (shown in Figure 6.9).

Stage one of the use of a hydrological simulation model involves the provision of initial information about the drainage-basin. Such information is obtained from maps and from hydrological and meteorological records. Stage two is termed *calibration*. It involves the testing of the performance of the simulation model using known events. Given an initial set of conditions the model is run in order to see how closely the real and simulated events correspond. Various parameters of the mathematical model, which represents the relationships between system components, are modified until real-world and simulation model behaviour agree as closely as possible. Once this stage has been completed successfully then stage three is implemented. At this stage the model is used to predict future events, using simulated sequences of precipitation that bear a close statistical resemblance to real sequences. The statistical resemblance is measured by comparing mean values and ranges of actual and simulated sequences, as well as looking at the distribution of runs of wet or dry days.

The results from the model can be used to answer questions relating to the long-term behaviour of the river basin. For example, given a suitable calibrated river basin simulation program a sequence of 100 or 200 years of daily observation can be generated and estimates made of the flow level that is exceeded on average once every 10, 50 or 100 years. These levels are called the 10-year flood, the 50-year flood and the 100-year flood, respectively, and they are used by engineers and planners in the design of structures such as bridges and dams. Another use of the simulation model might be to assess the effect of increasing the urban area lying within a river basin. Precipitation is conducted to the nearest water course much more quickly in an urban area than it is in a natural drainage basin, so the building of a new town or the extension of an existing town can have a considerable effect on the rate at which the water level of a river rises after a storm. The rate of increase of river flow will be greater in the urbanized basin than it would be if the basin were still covered by natural vegetation. In both these examples it is clear that a major advantage of the simulation model is that answers can be more reliably obtained to questions such as "how high should I build this bridge?" or "what would be the hydrological effect of building another 2000 houses in Edwalton?" than if the engineer or planner were to rely solely upon rules of thumb or "common sense".

Another example of the use of simulation models is in the estimation of the yield of a drainage-basin. Rivers, together with groundwater reserves, supply much of the water used for domestic, industrial and agricultural purposes. Overuse can lead to pollution, declining groundwater levels and, eventually, to trouble of one sort or another; internecine feuds were reported in southern Spain in the summer of 1983 when river levels were at an all-time low. For the sensible management of water resources, good estimates of the total amount of water available, on average, per year together with a measure of the variability in supply from year to year if supply and demand are to be kept in balance. Such estimates can be made readily if a basin simulation model is used.

6.7 SUMMARY

Simulation is the representation of a system in terms of another medium or at another scale. A simulation model could be, for instance, a physical model such as a wave-tank model of a beach. Digital or computer simulation uses the medium of mathematics to represent a system and its behaviour. Mathematical expressions are used to represent real-world relationships. The advantages of computer simulation are:

- speed of execution – 100 years in the life of a watershed can be simulated in a few seconds of computer time, and
- flexibility, which implies that the computer simulation model can be altered easily to take account of changes in circumstances or to allow modelling of a different system. Experiments using the simulation model are thus made possible; an example given earlier showed how a planner could modify a river-basin simulation model to allow for the expansion of an urban area. Such flexibility is not possible with a physical model.

Computer simulation models which incorporate random effects have been described in this chapter. This kind of model is termed a stochastic model, and examples include random-walk models of stream patterns and river-basin simulation models which use randomly-generated rainfall sequences. Other simulation models, such as population projection models, are usually deterministic. A random element is incorporated into stochastic models to summarize or act in place of a number of small, independent processes. No matter whether the stochastic or a deterministic form is used, the successful use of the model will depend on two factors, which are

- the degree of understanding of the operation of the system that we possess, and,
- the quality of the data used to calibrate the model.

The first factor helps to explain why simulations of physical systems are often more successful than simulations of social, demographic or economic systems, for physical systems are generally easier to isolate, identify and characterize. In addition, human systems may well respond or adapt to the predictions of a model, which become, in effect, self-fulfilling prophecies. If the Treasury economic simulation model predicts 15% inflation in 18 months' time then people's behaviour may well be influenced by the expectation that the model is correct in its predictions, no matter how good or objective the data on which the simulation was based.

6.8   REVIEW QUESTIONS

1. Define:

random walk	random number	Markov chain
population cohort	closed system	open system
hydrological system	calibration of model	stochastic model
deterministic model		

2. Describe the types of simulation model generally used by geographers. In what way do they differ?
3. Summarize the factors which make simulation models necessary to and useful in geographical studies.
4. Outline the problems met with in designing a model to simulate changes in the size and structure of a human population.

CHAPTER 7

Geographical Information Systems

7.1 INTRODUCTION

A widely-accepted model of the relationships between the Earth's atmosphere, land and oceans and their component parts is that of an open system made up of interrelated subsystems (Chapter 6; Chorley and Haggett, 1969). The boundaries between these subsystems, such as the hydrological and the atmospheric subsystems, are not sharp or clear. Furthermore, our perceptions of the behaviour of these subsystems depend strongly on the scale at which they are observed. Subsystems can, for ease of human understanding, be labelled as physical (such as geology, hydrology, climate, soils, vegetation and fauna), economic (trade, transport, industry), political (organization, population size and distribution) and so on.

Much of geography over the last 40 years has been concerned with analysing these subsystems by breaking down each subsystem into ever smaller component parts and studying these parts, and their interrelationships, at an increasing level of detail. On the other hand, decisions about real-world problems involve the integration or synthesis of these various subsystems, and the appreciation of interactions between them. Perhaps the most topical example concerns the relationship between global-scale systems of vegetation, climate, oceans and human activity which are thought to influence, or at least accelerate, processes of global environmental change. In a more localized example, a county planning department would need to consider a wide range of environmental, economic, political, social and spatial factors when deciding on a site for a public service utility such as a thermal electricity generating station. A distinction can therefore be made between approaches to the study and application of geography that involve the analysis of systems on the one hand and the synthesis of facts and knowledge concerning sets of systems and related to particular places on the other. These approaches have traditionally been termed systematic and regional geography. Earlier chapters of this book have been concerned with the analysis and display of geographical data. This final chapter looks at ways of putting it all together. Readers will therefore find it helpful if this chapter is read last.

Computer systems for the integrated handling of geographical or spatial data are called *Geographical Information Systems* or *GIS* for short.

The most commonly-used device used by geographers to integrate data relating to the physical, social, economic and political characteristics of an area is the map. A map is both a repository of facts and a tool for drawing inferences. The first of these properties is readily appreciated by looking at an atlas. Questions such as "what is the name of the river on which New Orleans stands?" or "what is the height of Ben Nevis?" can be answered by referring to the appropriate page of the atlas. The second property, that of allowing inferences to be made, requires a map showing two or more features of an area, such as the distribution of tree species and height above sea-level. The human eye and brain can produce generalizations such as "in this particular area, pine trees grow only at heights greater than 90 metres above sea-level". This type of generalization requires the comparison of maps showing the two spatial patterns of elevation of the terrain above sea-level and the distribution of pine trees over that same terrain.

It is evident from this discussion that a map contains two distinct types of information. One type refers to the geographical location (both absolute and relative) of spatial entities (the points, lines and polygons considered earlier in Chapter 2) and the other data type refers to the properties or attributes of such spatial entities, for example the height of a point such as a hill top, the width of a road (line) or the area of a State (polygon) Table 7.1). In drawing the inference above, concerning the relationship between the occurrence of pine trees and height above sea-level, we used (i) the location of individual points and (ii) the two properties of those points, the first being a continuous variable (height above sea-level) and the second a binary or presence–absence variable (pine trees/no pine trees). Queries that can be answered by a map can refer to either or both types of data. One type of query refers solely to the absolute or relative locational properties of the spatial entities. Some examples are:

- Given a network of lines representing highways, what is the shortest route between any two points in the network?
- How many of a set of points at which rainfall is measured lie within 50 km of Cambridge (relative location) or in the rectangle defined by the lines of latitude 55°N and 58°N and the lines of longitude 1°W and 3°W (absolute location)!
- Show the region lying within 5 km of the M1 motorway.

Other queries may relate to specific attributes of the spatial entities, or to combinations of these attributes, to the exclusion of any locational property. Examples of this kind of query include: How many states of the United States have a population greater than two million people? What percentage of the area of New Jersey is classified as urban? What is the population density (persons per square kilometre) of Alaska?

Table 7.1   Examples of point, line and polygon spatial entities and associated attribute data

Spatial entity	Description	Typical attributes
Point	raingauge	hourly rainfall readings
	trigonometrical point	height above sea-level
	river flow station	continuous streamflow record
	motorway junction	junction number
Line	road	road number
		classification (motorway, trunk, A, B, or unclassified)
		description (dual carriageway, restricted access)
	river	name
		width
		flow statistics
Polygon	State	name
		population
		time zone
		average annual rainfall
	catchment	slope
		maximum stream order
		percentage vegetation cover
	census tract	total population
		number of persons over 65
		number of households
		percentage households owning more than two cars

The third kind of request relates both the locational properties of spatial entities and their attributes: for example, pick out those regions of type X which have among their neighbours at least one region of type Y. List those states of the USA which have more than 1000 kilometres of Interstate highway. Find all towns in the state of Texas that have a population of more than 10 000 and are more than 50 kilometres from a railway.

These types of questions cannot be answered easily by looking at a conventional map. First of all, no map can show all the attributes that may conceivably be required by a user. Two or more maps would be needed if the detail shown is not to get too great to be displayed or absorbed by the user. Thus, in order to answer a particular query maps showing geology, soils, transportation, elevation, slope, drainage lines, vegetation types, populations of towns and planning designations may be required. If the area of interest is not trivially small then it would not be possible to show all this information on a single map. We would thus have to resort to the use of overlays, with one overlay per attribute. If each overlay were drawn on transparent paper then a light table could be used to allow the viewer to look simultaneously at the spatial distribution of each attribute and to pick out by eye the areas of interest.

The second reason why conventional maps are unsuitable for the answering of the types of queries mentioned above is that the human eye and brain cannot take in the amounts of information that would be present in the overlays and in the accompanying tables of figures and statistics. The volume of information that would have to be absorbed is simply too great. If the cartographic and tabular data could be stored in computer-readable form, then the speed of the computer could be used to select areas with the required characteristics, derive composite maps, and perform other operations that would be impossibly time-consuming if carried out by hand. A Geographical Information System provides just such capabilities. A GIS can be represented as a means of producing digital overlays of spatially-referenced information, together with the means of entering, recovering, processing and displaying information present in these overlays (Figure 7.1). It has been estimated that over 1000 GIS are in use in the United States. This number is expected to grow to 4000 over the next five years. Several private-sector companies have specialized in the production of GIS; among the best known are Arc-Info (ESRI, USA), Intergraph (Intergraph, USA), Informap (Synoptics, USA), I²S (International Imaging Systems, USA), SPANS (Tydac Technology Ltd, Canada), Sicad (Siemens, Germany), METROPOLIS (LaserScan Ltd., UK)

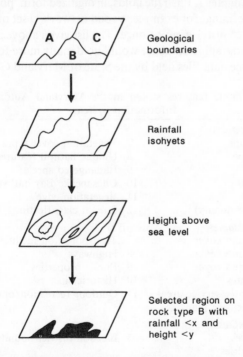

Figure 7.1 Conceptual structure of a geographical information system: maps of various features are converted to a common scale and projection so as to allow the identification of regions that satisfy particular requirements.

and System 9 (Prime Computers, USA and Wild Heerbrugg, Switzerland). (These products are mentioned for information only; such mention should not necessarily be taken as a recommendation.) Geographical Information Systems that can run on a personal computer are now becoming available; examples are SPANS and PC-ARC/INFO. All of these systems have several characteristics in common, namely data entry and editing, data manipulation, data display and modelling. These features are considered in turn in the following sections.

7.2   DATA ENTRY AND EDITING

The data stored and manipulated within a GIS are of two specific kinds. The first is cartographic and defines the relative or absolute location of spatial entities (points, lines and polygons) in the area of interest. The second kind of data consists of the attributes or properties of the spatial entities shown on the maps.

Cartographic data consist of digitized maps, that is, maps which have been converted to numerical form by the process of digitizing, which was described in Chapter 4. Each feature of a conventional topographic or thematic map is stored separately in a cartographic database, which consists of a number of files. Files were described in Chapter 1. Each file holds, in digitized form, point, line and polygon information defining, for example, a road network, a set of soil boundaries, river lines, or sites of historic importance. The list given above is just a selection; an operational cartographic database would hold many more features. Table 7.2 lists the cartographic data files held by the State of Maryland GIS (the Maryland

Table 7.2   Cartographic features stored in the Maryland Automated Geographic Information System

I: Physical data variables		
1 Natural soil groups	7	Vegetation cover types
2 Topographic slope	8	Unique natural features and scenic areas
3 Geology	9	Endangered species
4 Mineral resources	10	Chesapeake Bay bathymetry
5 Wetlands	11	Vegetation edge
6 Hydrology (water quality)	12	Stream classification
II: Cultural data variables		
1 1970 Land use/land cover	7	Transportation facilities (non-highway)
2 1973 Land use/land cover	8	Highways
3 1978 Land use/land cover	9	Public properties
4 Archaeological sites	10	Historic sites
5 County sewer/water service areas	11	Outdoor recreation/open space
6 County comprehensive plans		
III: Areal data variables		
1 County boundaries	3	Election district boundaries
2 Watershed boundries		

(*Source*: *MAGI: Maryland Automated Geographic Information System*, Maryland Department of State Planning, Publication #81-13, revised April 1981, Baltimore, Maryland. Reproduced by permission.)

Automated Geographic Information System). The importance of the file-structured database lies in the fact that the contents of the individual files can be abstracted and combined to serve a particular need. The way the database is structured (Chapter 2) will, however, determine the types of question that the information system can be asked.

In Chapter 4 a brief account of the process of map digitization was given. The manual procedure involves following lines (such as contour lines or election district boundary lines) with a pen or stylus. An intelligent controller can track the position of the pen and record its location either at a fixed rate (such as 16 times per second) or whenever the operator presses a button. More expensive systems use laser beams to follow lines. A third kind of digitizer is becoming more popular. This is the scanning digitizer which uses an advanced type of TV camera to convert the map to raster or cell format (Chapter 5). This is the pixel format in which remotely-sensed images are stored. The raster-coded image, in its simplest form, has the value '1' wherever the cell lies over a line and the value '0' otherwise. Scanning has to be followed by computer processing to join up, follow and extract the lines in vector format for storage in a cartographic database. Figure 7.2 shows

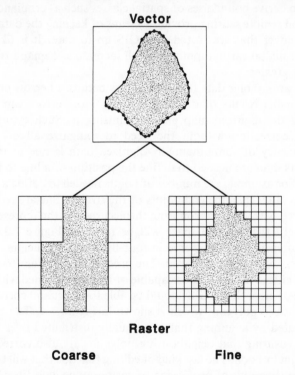

Figure 7.2   Raster and vector representation of locational data. The vector data in the upper figure are stored as a string of (x,y) coordinates. The correspondence between the vector and raster representations depends upon the fineness of the raster grid (lower figure).

the distinction between raster and vector representation, and further discussion can be found in Chapter 2. The choice of method of digitizing depends more often than not on the level of funding available. Manual digitizing is slow but is relatively accurate and cheap. Laser scanning is fast and expensive. Raster scanning/vectorization is also fast, and involves almost no user intervention. However, the cost is high (but falling) and the quality of the final product is sometimes not up to the standard of high-class manually digitized material.

An increasing amount of cartographic data is becoming available in digital form. In the United States the US Geological Survey already provides digital elevation models (DEMs) and digital line graphs (DLGs). The latter were described in Chapter 2. A DEM is a grid of terrain elevations which can be converted to vector (contour line) form. Digital maps of slope and aspect can be derived from a DEM if a suitable computer program is available. The Ordnance Survey of Great Britain is also providing a limited amount of digital map data and DEMs. Processed remotely-sensed images are another source of machine-readable cartographic data. A classified Landsat TM image (Chapter 5) can be used to derive boundaries of spatial classes such as cropland, woodland and water, and remote sensing provide a means of keeping the data concerning land-surface cover that are stored in a GIS up to date. It is likely that the availability of digital cartographic data will increase as demand from users of GIS becomes greater.

Editing of cartographic data files is necessary because of errors or deficiencies in input or in the results of data processing. Gross errors can be detected by displaying the digitized map and scrutinizing it. Such errors should be amenable to correction without the need to redigitize all or part of the map. A deficiency of some manually-digitized data is due to the inclusion of more points than are needed to define the position of a line to a given level of accuracy. For example, the number of points needed to define a straight line is two. Points other than the end points of the straight line are redundant and can be removed without compromising the accuracy of the representation and with the benefit that storage space will be reduced (Figure 7.3). If the x,y coordinates of each point are stored in four-byte units then eight bytes will be saved for each excess point removed. The process of checking for unnecessary points on a digitized line should be capable of being performed within the GIS. Another type of error that a GIS should be able to detect and correct is closure error in a digitized polygon, which is shown in Figure 7.4. The closure error is usually treated by assuming that it is equally distributed over the polygon nodes whilst ensuring that neighbouring polygons are also corrected for any changes in point location. The last kind of editing function that will be considered here relates to the result of processing of cartographic data files. The overlay of two sets of polygons representing (for example) soil and geological boundaries respectively may result in discrepancies in the form of small "sliver" polygons

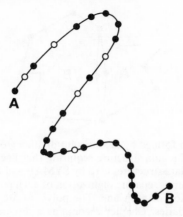

Figure 7.3   Editing of input data. Elimination of the points shown by open circles does not compromise the accuracy to which the line can be represented on a digital plotter, yet computer storage and processing times are reduced. The number of points needed to represent the line will depend upon the scale of the map; there is no universally-valid number.

(Figure 7.5). Such slivers may be an accurate representation of the true situation but it is more likely that they are the result of inaccurate digitization, imprecise surveying, or the use of maps of considerably different scale. A GIS should be able to identify and remove slivers, and correct the resulting map appropriately. However, error resulting from map generalization cannot be removed automatically.

Attribute data relate to the properties of the points, lines and polygons that are stored in the cartographic database. The attribute data are held in a separate database. Each attribute or set of related attributes is stored in its own file. One such file might hold demographic attributes for the individual counties of a given state; these attributes would include total population and population

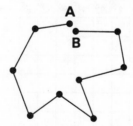

Figure 7.4   Closure error in polygon digitizing. Points A and B should be the same, but operator or system errors have resulted in a discrepancy. The error can spread over the polygon by altering the positions of the boundary points appropriately. However, it must be remembered that these points also lie on the boundaries of adjacent polygons.

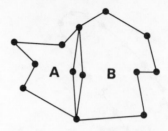

Figure 7.5   Sliver polygon formed by inconsistent digitizing of the boundary between polygons A and B. Use of a data structure requiring that lines be digitized once only is preferred; compare the data structures used by SYMAP and GIMMS in the examples in Chapter 4. SYMAP requires separate digitization of each polygon, whereas GIMMS requires separate digitization of each line. The polygons are then assembled from a description of their boundaries. Chapter 2 contains a discussion of alternative data structures for spatial data.

broken down by age and sex while a second attribute file would store the mortality (death) and morbidity (sickness) rates for the same counties. These attribute files can be envisaged as tables. The rows of each table refer to the spatial entities, whether they be lines, points or polygons, while the columns refer to the variables that are measured on these spatial entities (Figure 7.6). Attribute data is most often obtained in computer-readable form direct from agencies whose function it is to collect and distribute such data. The best-known example is the census of population. Other sources of attribute data are published maps (such as maps of soils and geology) and publications associated with such maps.

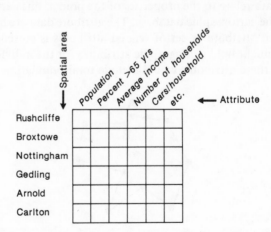

Figure 7.6   Example of table holding attribute data for a set of areas. The table is stored as a disc file, with each row of the table being held as a record (see Chapter 1). Sets of related files are managed and organized by a Data Base Management System (DBMS).

The problems associated with data collection, input and editing should not be underestimated. Indeed, it is possible to speculate that the progress of GIS is not being held up by lack of technology or of suitable software but by the cost of acquiring the data necessary for the functioning of the GIS. Typically, the cost of acquiring, checking, editing, and converting data may be of the order of the system hardware and software costs while the time required to assemble and check the cartographic and attribute databases may be of the order of several years. The need for data is not restricted to this start-up period, for new data are being generated continuously as roads and towns are built, and as the population changes both in size and constitution (due to births, deaths and disease) and spatial location (due to migration). Perhaps the greatest mistake in the application of GIS has been to allow the glamour of the technology to distract attention from the more mundane, yet vital, aspects of data acquisition and verification.

7.3   DATA MANIPULATION

The major attraction of a fully-fledged GIS is that it gives access to large volumes of cartographic and attribute data which can be manipulated according to the needs of the user and the flexibility of the system software. An overview of the manipulative functions offered by a GIS for both cartographic and attribute databases is given in this section.

Cartographic data manipulation operations include

- transformation of data from one map projection or scale to another,
- joining of adjacent maps,
- polygon extraction,
- polygon overlay and selection,
- buffer generation,
- network analysis, and
- redistricting.

These functions are considered in turn.

Transformation from one map projection to another will be necessary where two or more maps to be overlaid are drawn on different projections or on different scales. Snyder (1982, Appendix B) lists 20 map projections used by the US Geological Survey, and individual states make use of their own projections and grids. The same is true of the individual countries making up the European Community. It is also not uncommon to find that the projection used to map a particular phenomenon is altered from time to time. Hence, the ability of a GIS to transform map data quickly from one map projection to another is essential if multi-source and multi-temporal data are used. The topic of map projections is covered briefly in Chapter 4, while the conversion of satellite imagery to a cartographic coordinate system is considered in Chapter 5.

Thematic maps from satellite imagery represent an important source of data, particularly in third-world countries.

Captain Edward Aloysius Murphy is alleged to have said in 1949 that if a way exists to do a job wrongly, one day someone will do it that way. This observation is applicable to GIS in a modified form. A spatial area of interest to a geographer will, if possible, lie in the corners of four adjacent map sheets, thus increasing the cost of acquiring maps and maximizing the inconvenience involved in using the maps. A fully-operational GIS should be able to extract an area of interest, given the coordinates of two opposite corners of a rectangular region containing the area, irrespective of the number of map sheets that the region occupies. The map sections should be joined together without the seams being visible. It may be the case that adjacent maps use different projections or scales. This can happen if a map series is being updated; for example the Ordnance Survey of Great Britain changed its standard scale from 1 : 63 360 (one inch to a mile) to 1 : 50 000 (one centimetre to 500 metres) during the 1970s. A GIS should be capable of handling this problem.

An area of interest is not necessarily rectangular; a GIS must be able to handle the general case in which the area of interest is polygonal. Examples are counties, states, islands, lakes or geological regions. If we wish to study Lake Tahoe from the point of view of hydrology, fauna and water quality we may find it necessary to extract from maps of these features the polygonal area defining the extent of the lake. The separate polygons showing the hydrological, biological and water quality parameters of Lake Tahoe should be capable of being overlaid. In some cases the overlay will not be completely accurate as the different maps may have been produced at different times by different teams of surveyors. Indeed, the outline of the lake may have altered in the times between the individual surveys. The result will be the generation of sliver polygons along the boundary of the lake. A GIS is capable of identifying such slivers and allowing the user the choice of leaving them as they are or of replacing them with an average boundary position.

If the spatial feature of interest is linear (such as a boundary, river, a railway or a pipeline) then the user of a GIS may wish to generate buffer zones on either side of the linear feature. These buffer zones could define regions of ecological interest (such as habitats defined by distance from a forest edge) or environmental (noise levels as a function of distance from a highway, aircraft flight path, or railway). Given the location of the linear feature the GIS is able to define the boundaries of the buffer zone and extract those geographical features of the zone that are required by a user.

A set of linear features forms a network. Examples are river and road networks. If the road network of an area is taken as an example one could envisage it as a set of nodes (representing junctions) and links (the section of road between two junctions). The nodes (junctions) may have attributes if they represent geographical entities such as towns and cities, and the links have the attributes of distance, traffic capacity, average speed sustainable by a car. Questions that are often asked about networks of this type are: what is the fastest route from A to B? Is it possible to travel through the network visiting each of a specified set of nodes only once? What is the most efficient way of travelling from A through B and C to D? If such questions can be answered by a GIS then considerable sums of money could be saved by companies involved in distribution and sales. A branch of mathematics called network analysis has developed to provide the answers to the type of problem just cited, and network analysis algorithms are incorporated into comprehensive Geographical Information Systems as well as being available for stand-alone PCs.

The final example of a cartographic operation that can be carried out by a GIS is redistricting. This term refers to the generation of a set of boundaries to define regions or areas which have a particular property or set of properties. Thus, electoral districts in the United States must have, as far as is possible, an equal number of voters, since all men (and women) are considered by the US Constitution as having been created equal. If the census data and the voters' register are available for each block in a particular city then electoral boundaries can be defined on the basis of equal voting populations per electoral district, and maps of these districts can be drawn. The UK practice is rather more subjective and takes into account socio-economic factors as well as absolute numbers of voters. However, constituency boundaries are changed from time to time. While perhaps improving the democratic process this redistricting procedure does mean that comparisons between electoral districts over time is difficult, since the boundaries of the units are not constant from one decade to the next.

The manipulation of tables of data such as the census data used in the example in the preceding paragraph is a problem not unique to geography. The rows of the table could refer to any set of objects (such as the cars produced by Rover) and the columns to the properties of these objects (for example the parts required in the manufacture of the cars). Questions such as "how many parts are unique to a single model?" could be asked. Most large commercial corporations and government agencies have assembled sets of tables of this kind. They are called databases. Computer manufacturers and software developers have invested large amounts in building computer programs to handle queries addressed to these databases. The computer programs are generally highly complex but are becoming more and more user-friendly as psychologists and computer scientists develop easier-to-use human–computer interfaces or "front-ends" using the concepts of artificial intelligence. Database Management Systems (DBMS) form a growing proportion of business in the software world. Several kinds of DBMS are in use, and the interested reader should refer to an introduction to database management systems for a guide to the different types. The review by Date (1986) is recommended.

The easiest type of DBMS to understand and use is the Relational Database Management System or RDBMS, which was mentioned earlier in this chapter. A RDBMS uses the concept of the table (with the traditional row–column layout) as its basic structural entity. Each table is stored as a file on the disc system of the computer (see Chapter 1) and tables can be searched for occurrences of events like "all spatial entities having attributes A and B but not C".

There are thus two types of query that are answered by a GIS. One relates to the absolute and relative locational properties of the spatial entities themselves (the points, lines and polygons) while the second refers to the attributes or properties of the spatial entities. Of course, the user of a GIS would not be concerned exclusively with relative or absolute location to the exclusion of the attributes of the spatial entities, nor would he or she be interested solely in the non-spatial characteristics (the attributes) of the spatial entities. However, in each particular phase of an enquiry, the primary concern would be one or the other of these two types of query. In order to be able to handle these queries the GIS has two database systems. One handles spatial queries, which refer to absolute locations or to spatial relationships (such as adjacency or contiguity). The latter type of relationship is sometimes called *topological*. The second DBMS handles queries concerning the attribute set. For example, the ARC/INFO GIS produced by ESRI uses its own ARC spatial database system and a third-party relational database management system called INFO. The two are interfaced together so that queries that are primarily spatial can also access the attribute data, and vice-versa.

7.4   DISPLAY OF DATA AND RESULTS

The techniques for displaying spatial data have been reviewed in previous chapters. These techniques can be divided into two kinds: those which produce a permanent paper or photographic (human-readable) output and those which generate a temporary display on a television monitor screen. Most large-scale GIS use the television monitor display in an interactive fashion, so that users can manipulate the scale, colouring and annotation of an output map until they are satisfied. At this point the map can be sent to a hardcopy device such as a graphplotter, an electrostatic printer, an inkjet or matrix printer (in descending order of quality and cost). These devices were described in Chapter 4. Temporary screen displays are generated on a high-resolution monochrome or (preferably) colour monitor with at least 640 phosphor dots in the horizontal direction and 400 dots in the vertical direction. The number of dots on the screen (in this case 640 by 400) defines the resolution of the screen. Higher resolution costs more money but if the resolution is too low then straight lines look jagged and circles become regular polygons. Some computer systems use raster graphics which generally have a lower resolution than the more expensive vector graphics. The distinction between vector and raster systems was made in Chapter 4.

Not all of the output from a GIS is in the form of maps. Provision must also be made for the output (in hardcopy form) of tables and graphs. Ideally a software interface will exist to connect the cartographic, tabular and graphical output from a GIS and the report-generation capabilities of a high-quality word-processor, thus allowing the results provided by the GIS to be incorporated directly into a document.

7.5 MODELLING

The world is not static. If it were, planning would be a very boring occupation. As it is, the greatest rewards accrue to those who, like Arthur C. Clarke, are most successful in forecasting events. Not all of us can aspire to this level but we can ask questions such as "what would be the effect of building a motorway between towns A and B?" Could we forecast the effect of the new road on traffic flows or on the price of adjacent land suitable for commercial or residential development? If we live in a free society we cannot carry out experiments on the real world so it would not be possible to go ahead and build the road just to see what would happen. Simulation techniques have been developed to allow scientists and planners to build working models of the systems which they are studying. Experiments can be carried out on these models rather than on the world represented by such models. The examples in Chapter 6 showed how these simulation models could be used in geography. To be used successfully a simulation model must adequately represent the system under study and there must be sufficient relevant data available to allow for the calibration or tuning of the model.

It should be apparent from the earlier sections of this chapter that a GIS provides both a database of spatial and attribute data and the software tools needed to manipulate and transform those data. A flexible GIS will also allow the user to experiment by building models of alternative future realities, such as the road development mentioned earlier, and to evaluate the outcome of such developments. As the volume of digital data becomes larger, and the pressures on the environment increase, so the use of a GIS to evaluate alternative planning strategies will become more cost-effective.

7.6 EXAMPLES OF GIS APPLICATIONS

7.6.1 Land use and environment

One way of understanding the world is to see it in terms of the relationships within and between a set of interrelated systems, as described in the opening paragraph of this chapter. This approach is becoming more widely accepted; for example, the following statement appears in the Report of the US Committee on Earth Sciences (1989, p. 4): "environmental changes are the result of complex interplays among a number of natural and human-related systems. For example, changes in the Earth's climate involve not only winds and clouds in the atmosphere, but also the interactive effect of the biosphere, ocean currents, human influences on atmospheric chemistry, the Earth's orbital characteristics, the reflective properties of the planet, and the distribution of water between atmosphere, hydrosphere and cryosphere''.

The development of a GIS capable of storing and manipulating the volumes of data required for a global research programme is still some way off, but smaller GIS have been used to attempt to provide at least a beginning. One such is an experimental GIS which has been developed for the CORINE (Coordinated Information on the European Environment) programme of the European Community. This program was begun in 1985 to harmonize the collection of data on the state of the environment in the European Community in order to improve environmental policy formulation and implementation (Wyatt, Briggs and Mounsey, 1988). Before an integrated environmental database could be established, it was necessary firstly to define standards for recording environmental information, secondly to develop modelling procedures to give predictive capabilities, and thirdly to specify appropriate hardware and software. Since the programme lasted only four years (1985–1989) and had a modest budget a limited number of priority areas was selected as follows:

- ecological sites of major importance for nature conservation;
- acid deposition, especially in relation to atmospheric emissions and potential damage to ecological sites and the soil;
- resources and problems of the Mediterranean region, including land use, land quality, water resources, problems of coastal regions and seismic risks.

In order to provide sufficient information on the three principal themes, data on around 200 individual environmental variables are in process of collection and recording. In addition to this attribute database a cartographic database is also being developed. In this cartographic database will be digital maps of coastlines, hydrography, slope, aspect, settlements, and administrative boundaries. In all, these maps correspond to 10–12 overlays. At a 1 : 1 000 000 scale each overlay consists of 200 Mb of data, a figure that is expected to increase to 500 Mb when the digitization programme is completed in 1989. A further 25 Mb of data at a 1 : 3 000 000 scale will also be stored.

The experimental GIS is mounted on a DEC VAX 11/750 running ARC/INFO software at Birkbeck College, University of London. A second workstation, running the Siemens SICAD software, is installed in the CORINE office in Brussels, and a third system, used for collecting land-cover data from remotely-sensed data for the territory of Portugal, is based on an I^2S image processor.

The CORINE GIS is still under development, and problems of data checking and verification will occupy a considerable amount of time before the GIS will be capable of producing reliable information. It is, however, in regular use by the Environment and other Directorates of the European Community and other collaborating national and international agencies. As it is one of the first international environmental GIS, the CORINE experience will be valuable in stimulating other, similar, projects and in providing experience on which other projects can draw.

A second international programme that is worthy of mention is the United Nations Environment Programme (UNEP), which was established in 1983 to coordinate global environmental management. One of the first environmental hazards to be studied was that of desertification. Key variables that were considered to affect the rate of desertification were soil status, vulnerability of land to desertification, and human and animal population pressures. Maps showing the distribution of these variables at a scale were digitized and desertification models developed by the United Nations Food and Agriculture Organization (FAO) were used within the GIS to produce hazard maps and summary tables. A second stage of the UNEP project was to start the Global Resource Information Database (GRID). The African continent was selected for the pilot project, and in 1987 a national environmental database for Uganda was established, using existing data and information derived from remotely-sensed images. This database was used in modelling crop suitability and erosion potential. Beginning in early 1988 a second phase was begun, and remotely-sensed data were selected as the source for databases on global forest resources and land degradation (Aronoff, 1989; Mooneyhan, 1987).

7.6.2 Social, economic and demographic

The South East Regional Research Laboratory (SERRL) of the UK Economic and Social Research Council (ESRC) is based at Birkbeck College and the London School of Economics, both colleges of the University of London. The most important aspect of the work of SERRL to date is centred on the building and maintenance of an integrated settlement and infrastructure database for South East England (Shepherd and Conway, 1988). The term *integrated* reflects the fact that the information in the database is referenced by its geographical location, so that maps of any part of the region can quickly be accessed. This is not the only advantage offered by an integrated spatial database for, given appropriate GIS software and hardware that are jointly capable of providing the kind of operations described in the earlier part of this chapter, geographical referencing opens up the possibility of combining cartographic and attribute data in logical ways, such as in determining those areas which possess characteristic A and (or/not) characteristic B.

The database holds 12 layers of cartographic layers, digitized from Ordnance Survey 1 : 50 000 map sheets. The main layers are:

- settlement;
- transport networks, including roads, surface and underground railways, and some utility distribution networks;
- administrative boundaries (wards, districts, counties, and electoral constituencies);
- planning areas (green belts, areas of outstanding natural beauty, development corporations, and sites of special scientific interest).

In addition, quantitative information from the 1971 and 1981 censuses and from other sources describing networks, areas and other infrastructure items such as air terminals, ports, bridges and tunnels is stored. Over 250 Mb of storage are required for the database, which is accessed by an ARC/INFO geographical information system running on a DEC VAX 11/750 computer. "The separate elements of the database add up to the essential spatial framework, in a co-ordinated and computerized format, for carrying out academic, commercial and policy-oriented studies in one of Europe's fastest growing regions" (Shepherd and Conway, 1988, p. 14).

Some examples of the use to which the SERRL database has been put are: the examination of the accessibility of small- and medium-sized towns to the motorway and trunk road network in promoting population growth, the analysis of the hinterlands of each of the 900 stations on British Railways' South Eastern Region, and updating the urban/rural boundary lines using remotely-sensed data. The population growth study took into account both accessibility and the impact of planning policies, including constraints such as green belts. The GIS gave rapid and easy access to the data in a number of different ways, it provided analytical tools for the production of statistical tables, and it allowed the output of maps and specific factors, such as the overlay of the road network onto the green belt layer, showing areas of high and low accessibility, thus permitting the evaluation of population growth potential relative to accessibility and planning constraints.

The second study used the transport layer of the database to identify the locations of 900 stations on the British Railways network. Circles of radius one, two, five and ten kilometres were drawn around each station and the census enumeration district data were used to produce maps and tables showing population numbers, structure and characteristics around each of the stations' catchment areas. Remotely-sensed data from the French SPOT satellite (Chapter 5) are being used in the third study to augment the cartographic data layers. Remotely-sensed data are acquired at regular time intervals, and thematic information such as land use can be obtained from such data using the techniques of classification described in Section 5.6.4. The urban–rural boundary can be monitored effectively and the cartographic database updated regularly using this combination of data and image processing.

7.6.3 Planning

A particular east-coast state of the United States, which has several large population centres, requires an assessment of potential sites for new major public facilities, in particular coal-fired electricity generating stations. The criteria governing the location of such facilities include:

(i) Environmental
- Cooling water shall cause minimum disturbance to the aquatic ecosystems of rivers and estuaries;
- sulphur-bearing smoke shall not pass over urban areas;
- the chosen site must not interfere with or reduce the value of scenic and recreation areas, nor must the habitats of rare or otherwise important plant and animal species be disturbed;
- If possible the buildings and other structures associated with the facility should not be visually intrusive.

(ii) Economic
- The generating station must be close enough to major population centres to ensure a sufficient labour supply;
- transmission distance from the generating station to population centres must be as low as possible consistent with other criteria;
- it should not be built on high-grade agricultural land;
- transportation costs for fuel and other raw materials should be minimized. A site near to a railway is preferred.

(iii) Engineering
- The selected site should be flat and require a minimum of earth-moving and levelling;
- the probability of flooding must be less than a specified level;
- foundation material must be of sufficient strength to support the structures to be erected;
- the probability of strong winds and tornadoes must be below a specified level;
- access to the site must not involve special difficulties such as steep gradients on approach roads;
- the site must be accessible during the winter months.

This list of requirements is not exhaustive and it is more than likely that the executive committee in charge of the planning process would change its mind at least once during the selection procedure as enquiries proceeded.

To attempt a solution to the problem of finding a set of possible locations for the electricity generating plant using manual methods would be extremely time-consuming and very costly. It would involve the use of models of rivers and estuaries in order to predict the pathways taken by pollutants, the modelling of smoke diffusion in the atmosphere, the comparison of many types of map (soils, recreational facilities, areas of scenic or historical value, plant and animal habitat, slopes, meteorological hazards, and river flow probabilities), and the analysis of demographic and economic data. If it were possible to store all the different types of primary data in computer-readable form then it would become feasible to use the speed of the computer to carry out the required searches, comparisons, overlays and numerical modelling. It would then be possible to use computer mapping procedures (Chapter 4) to display the results of these computations either in the form of hard-copy output or as a picture on a graphics terminal.

7.7 SUMMARY

The synthesis of geographical facts relating to the locational properties of spatial entities and their associated attributes is a necessary counterbalance to analytical studies carried out in physical and human geography. A rough distinction can be made between, on the one hand, the systematic branches of geography (economic, social, political and demographic) which are concerned with the detailed study of subsystems at increasing levels of complexity and, on the other hand, the application of geography to real-world problems. These applications require the assembling of data and concepts from the different systematic branches of geography to produce an integrated picture of the demographic, social, environmental and economic aspects of a particular area. In the past these synthesizing studies have been difficult to perform due to the lack of a technology which allows the storing, retrieval, analysis, manipulation and display of large volumes of data relating to areal units and their properties. This technology is now at the stage of commercial development, and the product is a GIS.

A GIS stores both cartographic data (showing topography or individual themes such as soils or geology) and attribute data associated with the spatial entities (points, lines and polygons) that are represented on the maps. The locational or spatial aspects of geographical problems are handled by a spatial database management system while queries relating to the attribute data generally use a relational database management system. The two database management systems are closely integrated to allow the solution of questions which have both a locational and a non-spatial content. In addition, the cartographic data can be manipulated in a variety of ways including projection transformation, scale change, mosaicing of adjacent map sheets, polygon extraction, overlay of maps showing different themes and the delineation of buffer zones. The cartographic data must exist in digital format before they can be used within a GIS; although the amount of digital map data is growing larger a considerable amount of manual digitizing is needed in practice to provide the database necessary for a successful GIS. Display of results is achieved in one of two forms: hardcopy (such as a plotter-drawn map) or as a display on a television monitor screen. Most often the monitor display is used interactively as the user experiments with and manipulates the output map in order to achieve a close fit to his or her requirements, and only then is output obtained in hardcopy form.

The technology for commercial GIS is now available. Two factors are holding back the development of GIS applications. One, as we have noted several times in this chapter, is the availability of spatial data in computer-readable format. The second is the lack of personnel who are trained in geography (so as to understand the problems of spatial analysis) and who have the necessary skills to understand and appreciate the uses of computers in solving those problems. The second of these two factors is likely to be the more important constraint in the long run, for even at the end of the 1980s it is rare to find a graduate in geography who is first of all aware of the wide-ranging scope of the subject and secondly is familiar with the operations capable of being performed by a modern computer system.

One particular difficulty that will cause increasing problems for the GIS user is that of *data quality*. It is now possible for a GIS to incorporate databases from many sources, while in the near future developments in networks will mean that remote access to databases at institutions other than the users' own will become possible. Many GIS users assume, perhaps subconsciously, that data quality and reliability must be high because the technology used to manipulate them is sophisticated. This is not a wise conclusion; very few data sources – of either locational or attribute data – give any guide to the reliability of the product. Gross errors may be present, but are fortunately rare. What is more insidious is the lack of guidance on the levels of accuracy to be expected or, where aggregate data are supplied (for example, the sum or average of the values at a number of points over an area) statements about the variability of these values within the area are exceptional. The question of data quality is one which will attract considerably more attention in the future, once technological developments have been absorbed.

A second difficulty is that data requiring specialist knowledge are becoming available to non-specialists. Incorrect or misleading conclusions may be drawn by naive users of, for example, medical data. Tables of mortality and morbidity values for spatial areas are apparently simple to interpret; however, unless such data are adjusted for the age-structure of the populations of those areas, any conclusions drawn will remain questionable.

Further reading on this new and growing subject can be found in Aronoff (1989) and Burrough (1986).

7.8   REVIEW QUESTIONS

1. Explain the difference between the analysis and synthesis as applied to the study and uses of geography. Into which category would you place (a) economic geography and (b) a geographical information system?

2. What is meant by the following terms:

spatial entity	attribute	database
RDBMS	buffer	digitizer
sliver polygon	map projection	transformation
mosaicing	shortest-route problem	
redistricting	(network)	

3. How far do you agree with the statement that the development of GIS applications will be held up by (i) lack of digital cartographic data and (ii) lack of skilled personnel? Give reasons to support your argument.

4. Give three examples of display devices used by a GIS. Comment on their applicability to particular problems of your choice.

5. Are there any problems and pitfalls connected with data sources facing the user of a GIS? Give examples. How should such problems be tackled?

APPENDIX A
World Data Matrix – Description and Listing

List of variables in data matrix.

The variable name used in the example runs of SPSS in Chapter 3 is given in capital letters following the definition of the variable.

1. 1987 population in millions (POPN).
2. Area in thousands of square kilometres (AREA).
3. Annual rate of population growth in percentage, mid-1980s (POPCHNGE).
4. Urban population as percentage of total population, mid-1980s (PCURB).
5. Economically active population not engaged in agriculture as a percentage of the total economically-active population, 1986 (PCNONAG).
6. Natural resource score based upon availability of bioclimatic, fossil fuel and non-fuel mineral resources per inhabitant. The score is simplified and telescoped as follows in relation to the world average score, which has been set to equal 100: (NATRES)

- 0– 24: 1
- 25– 49: 2
- 50– 99: 3
- 100–199: 4
- 200–299: 5
- 300–399: 6
- 400–599: 7
- 600–799: 8
- 800–999: 9
- 1000 + : 10

7. Energy consumption in kilograms of coal equivalent per inhabitant, 1984 (ENERGY).
8. Steel consumption in kilograms per inhabitant, 1984 (STEEL).

9 Supplied dietary energy per head in relation to nutritional requirements, 1980–1982 (in other words, the amount of food consumed per head as a percentage of the theoretical requirement given physiological characteristics of the population, climate, etc. (FOOD).
10 Telephones in use per hundred inhabitants (TELE).
11 Newsprint consumption in kilograms per ten inhabitants, 1982 (NEWS).
12 Deaths of infants under the age of one year per thousand live births (INFMORT).
13 Population over eligible age able to read as a percentage of all such population (LITERACY). Eligible age is that age below which it is not considered realistic or normal to read; this is usually 7 years, but is higher in some countries.
14 Gross National Product in US dollars per inhabitant (GNP).

The data are intelligent estimates for the countries numbered under the following variables:

- steel (variable 1) for countries 3, 42, 61, 68, 72, 76, 79, 83, 84, 91, 94, 97;
- telephones (variable 10) for countries 3, 4, 8, 25, 37, 39, 68, 70, 72, 78, 83, 84, 85, 86, 87, 94, 97, 99, 100;
- newsprint (variable 11) for countries 42, 61, 76, 79, 80, 84, 86, 91, 94;
- Gross National Product (GNP) (variable 14) for countries 3, 12, 21, 26, 37, 39, 43, 44, 48, 49, 51, 53, 62, 71, 83, 94.

LIST OF COUNTRIES

1. China	19. Egypt	37. Romania
2. India	20. Turkey	38. Kenya
3. USSR	21. Iran	39. Korea Dem. Rep.
4. USA	22. Ethiopia	40. Peru
5. Indonesia	23. Rep. of Korea	41. Venezuela
6. Brazil	24. Spain	42. Nepal
7. Japan	25. Burma	43. Iraq
8. Nigeria	26. Poland	44. German D.R.
9. Bangladesh	27. South Africa	45. Sri Lanka
10. Pakistan	28. Zaire	46. Australia
11. Mexico	29. Argentina	47. Malaysia
12. Vietnam	30. Colombia	48. Uganda
13. Philippines	31. Canada	49. Czechoslovakia
14. German F.R.	32. Morocco	50. Saudi Arabia
15. Italy	33. Sudan	51. Mozambique
16. United Kingdom	34. Tanzania	52. Netherlands
17. France	35. Algeria	53. Afghanistan
18. Thailand	36. Yugoslavia	54. Ghana

55. Chile	71. Angola	87. Haiti
56. Syria	72. Somalia	88. Hong Kong
57. Ivory Coast	73. Austria	89. El Salvador
58. Madagascar	74. Tunisia	90. Denmark
59. Hungary	75. Malawi	91. Burundi
60. Portugal	76. Burkina Faso	92. Finland
61. Cameroon	77. Zambia	93. Honduras
62. Cuba	78. Senegal	94. Chad
63. Greece	79. Niger	95. Israel
64. Ecuador	80. Rwanda	96. Paraguay
65. Belgium	81. Switzerland	97. Benin
66. Zimbabwe	82. Dominican Rep.	98. Norway
67. Bulgaria	83. Kampuchea	99. Sierra Leone
68. Mali	84. Yemen Arab Rep.	100. Libya
69. Sweden	85. Bolivia	
70. Guatemala	86. Guinea	

Sources:
1. World Population Data Sheet 1987, Population Reference Bureau, 2213 M Street, Washington DC 20037, USA (for variables 1, 3, 4, 12, 14).
2. J. P. Cole (1985): *Geography of World Affairs*, Butterworth, London (for variable 6).
3. Food and Agriculture Organisation Yearbook, Rome, Italy (1985) (for variable 5).
4. *United Nations Statistical Yearbook* 1983/4, United Nations, New York (for variables 2, 7, 8, 10 and 11).
5. 1982 World's Children Data Sheet, Population Reference Bureau, 2213 M Street, Washington DC 20037, USA (for variable 13).
6. *The State of Food and Agriculture*, Food and Agriculture Organization, Rome, Italy, 1985 (for variable 9).

LISTING OF WORLD DATA MATRIX

Countries are listed across the rows and variables are listed down the columns. The columns are numbered in the same order as the definition of variables given above. The three-digit number at the start of each row is the country identifier given in the table above.

	1	2	3	4	5	6	7	8	9	10	11	12	13	14
001	1062.0	9561	1.3	32	28	2	664	55	106	1	12	61	60	310
002	800.0	3046	2.1	25	31	2	237	17	92	0	4	101	42	250
003	284.0	22402	0.9	65	82	5	5977	550	132	12	45	26	98	5000
004	244.0	9363	0.7	74	97	5	9577	479	138	76	441	11	99	16400
005	175.0	1492	2.1	22	48	2	263	16	109	0	6	88	68	530
006	142.0	8512	2.1	71	75	4	656	70	108	8	24	63	73	1640
007	122.0	370	0.6	76	92	1	3800	569	122	54	240	6	99	11330
008	109.0	924	2.8	28	33	2	231	7	104	1	4	124	30	760
009	107.0	143	2.7	13	29	1	53	4	83	0	3	140	31	150
010	105.0	804	2.9	28	48	1	215	7	97	1	4	125	29	380
011	81.9	1973	2.5	70	67	3	1714	113	126	9	64	50	84	2080
012	62.2	333	2.6	19	38	1	123	2	93	0	0	55	40	200
013	61.5	300	2.8	40	50	2	312	12	106	2	13	50	89	600
014	61.0	248	−0.2	85	95	3	5564	489	126	60	215	10	99	10940
015	57.4	301	0.1	72	90	2	3105	366	141	43	55	11	96	6520
016	56.8	244	0.2	90	97	3	4760	255	129	52	239	9	99	8390
017	55.6	547	0.4	73	93	2	3923	276	141	60	102	8	99	9550
018	53.6	514	2.1	17	34	3	414	35	104	1	19	57	88	830
019	51.9	1000	2.6	46	57	2	641	53	130	2	21	93	42	680
020	51.4	781	2.1	46	48	3	865	105	121	7	33	92	66	1130
021	50.4	1648	3.2	51	67	5	1328	86	120	5	5	113	43	1000
022	46.0	1184	2.3	10	23	3	20	2	95	0	1	152	6	110
023	42.1	98	1.4	65	69	1	1524	198	125	17	58	30	92	2180
024	39.0	505	0.5	91	86	3	2180	170	134	36	61	11	93	4360
025	38.8	678	2.1	24	50	3	78	1	109	1	2	103	69	190
026	37.8	312	0.8	60	76	4	4494	416	128	11	33	19	98	3000
027	34.3	1223	2.3	56	89	6	2606	130	117	14	39	72	57	2010
028	31.8	2345	3.1	34	33	5	63	3	96	0	0	103	58	170
029	31.5	2777	1.6	84	88	5	1718	88	127	10	50	35	95	2130
030	29.9	1138	2.1	65	68	3	857	37	109	7	35	48	83	1320
031	25.9	9976	0.8	76	96	9	9773	525	129	66	422	8	99	13670
032	24.4	444	2.5	43	60	4	296	37	109	1	2	90	29	610
033	23.5	2506	2.8	20	32	4	77	3	99	0	1	112	26	330
034	23.5	937	3.5	18	17	3	42	3	105	1	2	111	74	270
035	23.5	2382	3.2	43	76	5	707	89	110	3	5	81	41	2530
036	23.4	256	0.7	46	75	3	2484	217	143	13	18	29	87	2070
037	22.9	238	0.5	53	77	3	4558	503	126	7	27	26	96	2000
038	22.4	583	3.9	16	21	3	95	11	88	1	4	76	49	290
039	21.4	121	2.5	64	62	2	2685	323	131	1	1	33	90	1500
040	20.7	1285	2.5	69	63	5	640	25	91	3	80	94	80	960
041	18.3	912	2.7	76	88	5	3100	173	104	8	84	38	81	3110
042	17.8	141	2.5	7	7	2	17	1	91	0	0	112	19	160
043	17.0	435	3.3	68	76	5	535	86	116	4	6	80	50	1000
044	16.7	108	0.0	77	90	3	7600	536	142	21	96	10	99	6500
045	16.3	66	1.8	22	48	2	120	5	98	1	8	30	79	370
046	16.2	7695	0.8	86	94	10	6128	370	117	54	408	10	100	10840
047	16.1	333	2.4	32	64	4	875	117	118	8	54	30	69	2050
048	15.9	236	3.4	10	16	2	24	1	76	0	0	108	48	150
049	15.6	128	0.3	74	88	3	6197	700	142	23	49	14	99	6000
050	14.8	2150	3.1	72	59	8	3640	332	123	15	7	79	16	8860

	1	2	3	4	5	6	7	8	9	10	11	12	13	14
051	14.7	783	2.6	13	17	3	94	2	80	0	1	147	28	150
052	14.6	34	0.4	89	95	3	5854	295	134	59	300	8	99	9180
053	14.2	658	2.6	16	42	3	69	2	84	0	0	182	15	150
054	13.9	239	2.8	31	45	2	76	2	72	1	2	94	49	390
055	12.4	757	1.6	83	86	7	920	64	111	6	6	20	93	1440
056	11.3	185	3.8	49	75	2	902	69	125	6	4	59	54	1630
057	10.8	246	3.0	43	41	2	205	8	115	1	1	105	40	620
058	10.6	587	2.8	22	21	3	52	3	111	0	4	63	34	250
059	10.6	93	− 0.2	56	85	3	3790	303	133	13	62	20	98	4000
060	10.3	92	0.3	30	77	2	1307	118	127	17	32	17	81	1970
061	10.3	475	2.7	42	39	3	465	3	93	1	1	103	51	810
062	10.3	115	1.2	71	79	3	1467	58	126	5	34	17	95	1200
063	10.0	132	0.2	70	74	3	2243	149	143	36	54	14	90	3550
064	10.0	284	2.8	51	67	5	677	15	91	4	37	66	79	1160
065	9.9	31	0.0	95	98	1	4939	353	140	43	181	9	99	8450
066	9.4	391	3.5	24	30	4	451	101	91	3	25	76	71	650
067	9.0	111	0.1	66	88	3	5523	323	147	20	46	16	95	3500
068	8.4	1240	2.9	18	17	2	30	2	74	0	0	175	13	140
069	8.4	450	0.1	83	95	3	4703	439	119	89	351	7	99	11890
070	8.4	109	3.2	39	45	3	194	11	96	2	15	71	51	1240
071	8.0	1247	2.5	25	28	4	120	8	99	1	1	143	3	200
072	7.7	638	2.5	34	26	2	98	2	90	0	0	150	5	270
073	7.6	84	0.0	55	93	2	4007	257	133	48	200	11	99	9150
074	7.6	164	2.5	53	68	2	671	81	115	4	8	78	57	1220
075	7.4	118	3.2	12	22	2	43	1	96	1	0	157	36	170
076	7.3	274	2.8	8	14	2	29	1	81	0	1	146	11	140
077	7.1	753	3.5	43	29	4	339	2	90	1	2	84	69	400
078	7.1	196	2.8	36	20	2	169	9	99	2	7	131	22	370
079	7.0	1267	2.9	16	11	3	56	1	105	0	0	141	5	200
080	6.8	26	3.7	6	8	1	34	4	91	0	0	122	50	290
081	6.6	41	0.2	57	94	1	3733	350	129	81	330	7	99	16380
082	6.5	49	2.5	52	59	2	490	12	95	8	24	70	74	810
083	6.5	181	2.1	11	28	2	3	1	89	0	0	160	36	100
084	6.5	195	3.4	15	34	1	198	3	104	1	0	137	8	520
085	6.5	1099	2.6	48	56	5	338	3	89	2	12	127	68	470
086	6.4	246	2.4	22	22	2	80	4	81	1	1	153	9	320
087	6.2	28	2.3	26	32	1	53	5	84	1	1	107	28	350
088	5.6	1	0.9	92	99	0	1761	257	118	40	164	8	84	6220
089	5.3	21	2.6	43	63	1	174	4	92	2	26	65	66	710
090	5.1	43	− 0.1	84	94	2	4521	357	139	75	307	8	99	11240
091	5.0	28	2.9	5	8	1	17	4	102	0	0	119	23	240
092	4.9	337	0.3	60	91	4	5002	444	117	59	336	7	100	10870
093	4.7	112	3.1	40	42	3	225	9	96	1	15	69	63	730
094	4.6	1284	2.0	27	21	2	21	1	75	0	0	143	6	200
095	4.4	21	1.7	90	95	1	2329	130	118	37	87	12	88	4910
096	4.3	407	2.9	43	53	4	231	2	123	3	15	45	86	940
097	4.3	116	3.0	39	36	2	44	3	94	1	0	115	25	270
098	4.2	324	0.2	70	93	5	6570	365	124	62	245	8	99	13890
099	3.9	72	1.8	28	34	2	67	2	84	1	1	176	7	370
100	3.8	1760	3.0	76	86	8	3772	114	160	3	22	90	56	7500

APPENDIX B

Creating an SPSS Data File

The World Data Matrix, listed in Appendix A, is used in the examples given in Chapter 3. These examples assume that the data matrix has been read by the computer and stored on a SPSS data file. It is important to distinguish between the SPSS filename (the local name of the file within SPSS) and the name by which the operating system of the host computer knows the file. The FILE HANDLE command in the listing below makes a logical connection between the SPSS filename PMMSPSS and the name PMMLIB.PMMSPSS. The latter name is that by which the file is known to the VME operating system of the Nottingham University Cripps Computing Centre ICL 3900 mainframe computer, on which all the examples in Chapter 3 were run. Readers wishing to use the World Data Matrix at their own local computing centre will have to make changes to the FILE HANDLE specifications. The SPSS commands needed to create an SPSS data file on disc for subsequent manipulation are:

```
DATA LIST FREE / POPN AREA POPCHNGE PCURB PCNONAG
                 NATRES ENERGY STEEL FOOD TELE NEWS
                 INFMORT LITERACY GNP
N OF CASES 100
BEGIN DATA
⟨. . . The data matrix is entered here with the values for each country starting
a new row. The World Data Matrix is listed in Appendix A
END DATA                                                    . . . ⟩
LIST
FILE HANDLE PMMSPSS / NAME = 'PMMLIB.PMMSPSS'
SAVE OUTFILE = PMMSPSS
FINISH
```

Note the remarks above concerning the FILE HANDLE command. The name by which the file is known to SPSS is PMMSPSS (although individual readers will use their own identifying name) while the name of the file (as known to the Nottingham University ICL 3900 series computer) is PMMLIB.PMMSPSS. Readers are cautioned against using the FILE HANDLE command with the exact specification given above; please check with your computer centre to determine the correct parameters.

APPENDIX C

Critical Values of Student's t

p_1		0.100	0.050	0.025	0.01
p_2		0.200	0.100	0.050	0.02
(df)	1	3.078	6.314	12.706	31.821
	2	1.886	2.920	4.303	6.965
	3	1.638	2.353	3.182	4.541
	4	1.533	2.132	2.776	3.747
	5	1.476	2.015	2.571	3.365
	6	1.440	1.943	2.447	3.143
	7	1.415	1.895	2.365	2.998
	8	1.397	1.860	2.306	2.896
	9	1.383	1.833	2.262	2.821
	10	1.372	1.812	2.228	2.764
	12	1.356	1.782	2.179	2.681
	14	1.345	1.761	2.145	2.624
	16	1.337	1.746	2.120	2.583
	18	1.330	1.734	2.101	2.552
	20	1.325	1.725	2.086	2.528
	25	1.316	1.708	2.060	2.485
	30	1.310	1.697	2.042	2.457
	40	1.303	1.684	2.021	2.423
	50	1.299	1.676	2.009	2.403
	60	1.296	1.671	2.000	2.390
	70	1.294	1.667	1.994	2.381
	80	1.292	1.664	1.990	2.374
	90	1.291	1.662	1.987	2.368
	100	1.290	1.660	1.984	2.364

Tables of critical values for Student's t. The row entries are degrees of freedom. The column entries can be used either for a one-tailed test (p_1) or for a two-tailed test (p_2). The one-tailed test is used whenever the alternative hypothesis specifies the direction of difference (e.g. the mean of x is greater than the mean of y) whereas the two-tailed test is used when the direction of difference is left unspecified (e.g. the means of x and y are not equal). See Section 3.4.4 for more details.

(Taken from E. F. Federighi, Extended tables of the percentage points of Student's t distribution. *Journal of the American Statistical Association*, 1959, **54**, 683–688, by permission.)

Table of Random Numbers in the Range 0–9999

2130	9460	6394	0544	5162	1499	2432	5400	5607	6668	5936	8224
6851	8154	9116	8423	3523	1084	1598	3905	7021	5950	4714	5345
3045	0535	4883	5072	2851	1014	6403	6808	4309	8994	1067	4534
3428	6235	0576	3661	2246	0933	8207	1218	6226	6359	6787	4090
6225	2652	7200	5259	0536	3340	5121	0498	8713	8221	0121	1278
7201	3717	7965	7373	1920	7989	2950	4530	9190	0786	1555	5316
3127	4219	5811	8697	7509	3011	6323	5526	8159	7519	5320	0179
7168	7547	1172	9409	9061	7931	9598	5432	4403	8294	7964	0833
4882	4851	4271	7772	6371	1688	6100	1381	2485	5673	4453	3912
2239	0438	7213	1673	9914	0184	0454	6151	4704	7957	6735	8475
3281	7998	3404	8393	7054	8810	6498	6866	1723	4680	7680	2795
2313	6550	1276	0052	3563	9537	2396	6678	2183	9592	3597	4308
6183	1105	6363	6420	6929	0611	0593	9214	8898	9544	9235	3378
5540	2529	8627	0496	5942	8498	1714	6673	8508	1541	0208	5193
6883	3305	9805	8387	2860	7638	1015	0407	8891	8709	7523	6808
2354	4099	1426	6484	1960	7685	9417	4086	8994	8486	7333	0407
2719	3577	7518	1318	2864	0282	9059	6680	7215	8471	4411	3710
8929	5529	1976	4429	9224	2300	3956	7927	9212	4056	6586	0226
9246	2074	1381	3520	8131	2446	6627	8955	9333	7815	4102	6917
8032	4514	3846	1762	2533	8725	4385	4578	6965	8184	6830	4084
6300	8231	5760	5000	1407	1398	0579	0998	7826	6217	3415	0559
1047	0597	3081	8614	3400	2479	4016	0060	0640	1217	8785	8845
0356	3870	8191	4132	7401	1774	5270	6272	4464	6797	6109	3123
8399	1072	2697	0600	0385	8417	4990	7074	0287	4761	8710	1527
3826	4781	7250	7748	4524	7436	5127	6829	2089	0720	1489	8707
3755	4142	3229	2030	6554	0454	4413	7322	1798	6774	6363	5130
4867	7170	3291	8157	5015	0830	3741	0003	2775	0468	1593	5407
3813	4701	8072	2196	5043	5940	6895	6457	6347	8421	2431	5097
7014	6991	3296	4411	9862	1484	5220	1583	8859	0186	4734	4221

```
3194   0955   6474   9786   6345   5649   2102   6668   2265   8257   1721   7434
7668   6442   1160   5937   2793   6513   6825   3212   8271   6714   0909   6830
8223   4254   0633   1840   0683   1941   1594   7788   6573   9577   0225   2412
7384   0189   2428   4061   4728   5707   7662   5112   4249   6589   7064   1979
5370   4616   5565   4135   9784   1507   6802   1679   0776   8372   3233   4678
3364   5132   9591   6644   6706   0036   2577   2164   7061   7935   5416   3438
6154   3735   2353   9330   5664   7290   9782   7621   3935   0395   5143   1422
7595   9788   4039   5987   6181   8476   3631   0205   9151   0372   7474   1068
2309   0673   9786   5114   3276   7455   2832   6827   5842   4028   2226   6464
4409   2890   9001   9973   1614   4268   7080   5974   1229   8253   0003   8966
7914   1285   5644   9461   5652   5521   1933   1022   4207   4489   9662   0229
```

APPENDIX E

Coastline Data for Canada, Baffin Island and Greenland

Latitude and longitude data used in map projection examples in Chapter 4

The data should be read across the rows with latitude preceding longitude in the conventional manner.

GREENLAND

80 23	79 23	79 26	78 27	77 25	76 20
75 21	74 25	72 21	71.5 22	72 27	70.5 27
70 23	68 30	64 40	60 43	61 49	66 52
69 52	70.5 54	71 52	72 52	71.5 55	75 57
77 60	76 68	77 70	77.5 68	78 72	80 70

BAFFIN ISLAND

61 64	63 67	62.5 63	66.5 66.5	66 63	66.5 61
68 66	70 68	71.5 72	72 76	72 80	74 82
71 83	74 85	73 89	71.5 90	70 88	70 82
69.5 79	68.5 76	66 73	65.5 74.5	65 74	64.5 75
63.5 71.5	61 64				

CANADA AND ALASKA

45 70	46 67	46 64	49 64	49 68	50 68
50.5 62	51 56	54 56	55 58	56 62	60 65
59.5 68	60 70	61 71	62 73	62.5 77	61 78
60 78	58.5 79	57 77	55 79	51 78	50.5 80
55 84	56 90	57.5 92	60 95	61 94	62 93
64 95	64 90	65 88	66 91	66 87	66.5 85
66.5 83	67.5 83	69 83.5	68 85	67 87	68 90
70 92	71 92.5	72 94	73 92	74 90.5	74.5 92
74 96	72.5 95	71.5 95.5	71 98	70 96	67.5 94

66.5 95	67 98	66 103	68 109	66.5 109	67.5 111
67.5 116	68.5 116	70 125	71 129	70 131	69 136
70 142	70.5 146	70 151	71.5 157	70 164	69 167
66 163	65.5 170	65 168	65 164	62.5 165	60 168
59 164	59 160	53 165	58 158	59.5 155	60.5 150
60 145	58 138	58 135	54.5 132	53.5 130	50.5 128
50 122	49.5 124	45 123			

APPENDIX F

Names of African Countries (Zones) used in SYMAP and GIMMS Examples

1. MOROCCO
2. ALGERIA
3. TUNISIA
4. LIBYA
5. EGYPT
6. MAURITANIA
7. SENEGAL AND GAMBIA
8. GUINEA – BISSAU
9. GUINEA
10. SIERRA LEONE
11. LIBERIA
12. IVORY COAST
13. MALI
14. BURKINA FASO
15. GHANA
16. BENIN
17. NIGER
18. CHAD
19. NIGERIA
20. CAMEROUN
21. GABON
22. CONGO
23. CENTRAL AFRICAN REPUBLIC
24. SUDAN
26. SOMALIA
25. ETHIOPIA
28. KENYA
27. UGANDA
30. RWANDA
29. ZAIRE
32. TANZANIA
31. BURUNDI
34. MALAWI
33. ZAMBIA
36. NAMIBIA
35. ANGOLA
38. ZIMBABWE
37. TOGO
40. MADAGASCAR
39. MOZAMBIQUE
42. REPUBLIC OF SOUTH AFRICA
41. BOTSWANA

APPENDIX G

Data for SYMAP Example

This appendix contains data for the 42 African countries used in the SYMAP examples in Chapter 4. Each column should be read separately with the data in the right-hand columns following after the data in the left-hand columns.

A-CONFORMOLINES

		X						
1	A	070.	295.	MOROCCO			080.	330.
		080.	330.				070.	375.
		120.	335.		3	A	070.	440. TUNISIA
		140.	310.				070.	465.
		150.	250.				110.	455.
		175.	245.				120.	470.
		190.	240.				150.	440.
		185.	205.				115.	425.
		240.	180.				070.	440.
		230.	135.		4	A	120.	470. LIBYA
		140.	235.				130.	505.
		115.	245.				155.	555.
		070.	295.				120.	560.
2	A	070.	375.	ALGERIA			140.	600.
		070.	440.				255.	610.
		115.	425.				280.	600.
		150.	440.				240.	510.
		220.	440.				250.	495.
		235.	465.				235.	465.
		290.	375.				220.	440
		205.	280.				150.	440.
		175.	245.				120.	470.
		150.	250.		5	A	140.	600. EGYPT
		140.	310.				150.	645.
		120.	335.				235.	725.

	255.	705.			390.	185.	
	255.	610.		11 A	405.	195.	LIBERIA
	140.	600.			425.	220.	
6 A	175.	245.	MAURITANIA		460.	220.	
	205.	280.			430.	180.	
	230.	265.			405.	195.	
	320.	265.			140.	705.	
	325.	180.			175.	705.	
	300.	155.		12 A	390.	220.	IVORY
	300.	130.			390.	250.	COAST
	280.	145.			400.	280.	
	230.	135.			455.	280.	
	240.	180.			460.	220.	
	185.	205.			425.	220.	
	190.	240.			390.	220.	
	175.	245.		13 A	205.	280.	MALI
7 A	300.	130.	SENEGAL		290.	375.	
	300.	155.	&		330.	365.	
	325.	180.	GAMBIA		335.	330.	
	360.	185.			350.	275.	
	355.	180.			390.	250.	
	360.	125.			390.	220.	
	320.	115.			360.	215.	
	300.	130.			360.	185.	
8 A	355.	160.	GUINEA-		325.	180.	
	380.	140.	BISSAU		320.	265.	
	360.	125.			230.	265.	
	355.	160.			205.	280.	
9 A	355.	160.	GUINEA	14 A	335.	330.	BURKINA
	360.	185.			370.	345.	FASO
	360.	215.			380.	335.	
	390.	220.			385.	315.	
	425.	220.			380.	290.	
	405.	195.			400.	280.	
	390.	185.			390.	250.	
	400.	160.			350.	275.	
	380.	140.			335.	330.	
	355.	160.		15 A	380.	290.	GHANA
10 A	390.	185.	SIERRA		385.	315.	
	405.	195.	LEONE		445.	325.	
	430.	180.			455.	280.	
	400.	160.			400.	280.	

	380.	290.	
16 A	370.	345.	BENIN
	375.	365.	
	445.	350.	
	445.	340.	
	380.	335.	
	370.	345.	
17 A	235.	465.	NIGER
	250.	495.	
	320.	505.	
	360.	480.	
	355.	375.	
	375.	365.	
	370.	345.	
	335.	330.	
	330.	365.	
	290.	375.	
	235.	465.	
18 A	240.	510.	CHAD
	280.	600.	
	330.	600.	
	390.	590.	
	430.	505.	
	400.	505.	
	380.	490.	
	360.	480.	
	320.	505.	
	250.	495.	
	240.	510.	
19 A	355.	375.	NIGERIA
	360.	480.	
	380.	490.	
	440.	430.	
	465.	425.	
	465.	385.	
	445.	370.	
	445.	350.	
	375.	365.	
	355.	375.	
20 A	380.	490.	CAMEROUN
	400.	505.	
	430.	505.	

	460.	490.	
	480.	510.	
	490.	505.	
	490.	470.	
	490.	425.	
	465.	425.	
	440.	430.	
	380.	490.	
21 A	490.	425.	GABON
	490.	470.	
	545.	485.	
	565.	440.	
	540.	420.	
	490.	425.	
22 A	475.	540.	CONGO
	525.	525.	
	575.	485.	
	580.	445.	
	565.	440.	
	545.	485.	
	490.	470.	
	490.	505.	
	480.	510.	
	475.	540.	
23 A	430.	505.	CENTRAL
	390.	590.	AFRICAN
	460.	640.	REPUBLIC
	475.	540.	
	480.	510.	
	460.	490.	
	430.	505.	
24 A	235.	725.	SUDAN
	305.	770.	
	315.	750.	
	365.	750.	
	430.	720.	
	465.	750.	
	470.	720.	
	475.	690.	
	460.	640.	
	390.	590.	
	330.	600.	

	280.	600.			475.	690.	
	255.	610.			530.	670.	
	255.	705.			555.	660.	
	235.	725.			585.	670.	
25 A	305.	770.	ETHIOPIA		620.	680.	
	365.	830.			675.	655.	
	400.	830.			655.	600.	
	415.	890.			610.	570.	
	460.	855.			615.	520.	
	465.	815.			590.	510.	
	465.	750.			590.	455.	
	430.	720.			580.	445.	
	365.	750.			575.	485.	
	315.	750.			525.	525.	
	305.	770.			475.	540.	
26 A	365.	830.	SOMALIA		460.	640.	
	385.	850.		30 A	530.	670.	RWANDA
	370.	930.			530.	690.	
	415.	925.			555.	680.	
	485.	880.			555.	660.	
	535.	810.			530.	670.	
	465.	815.		31 A	555.	660.	BURUNDI
	460.	855.			555.	680.	
	415.	890.			585.	670.	
	400.	830.			555.	660.	
	365.	830.		32 A	530.	690.	TANZANIA
27 A	470.	720.	UGANDA		530.	710.	
	500.	740.			580.	785.	
	530.	710.			655.	795.	
	530.	690.			665.	725.	
	530.	670.			635.	720.	
	475.	690.			620.	680.	
	470.	720.			585.	670.	
28 A	465.	750.	KENYA		555.	680.	
	465.	815.			530.	690.	
	535.	810.		33 A	620.	680.	ZAMBIA
	580.	785.			635.	720.	
	530.	710.			690.	705.	
	500.	740.			710.	670.	
	470.	720.			740.	605.	
	465.	750.			735.	580.	
29 A	460.	640.	ZAIRE		715.	570.	

	680.	575.			795.	680.	
	675.	600.			790.	655.	
	655.	600.			740.	605.	
	675.	655.			710.	670.	
	620.	680.		39 A	655.	795.	MOZAMBIQUE
34 A	635.	720.	MALAWI		700.	800.	
	665.	725.			760.	725.	
	700.	740.			820.	730.	
	725.	730.			830.	695.	
	700.	720.			830.	685.	
	690.	705.			795.	680.	
	635.	720.			730.	705.	
35 A	590.	455.	ANGOLA		710.	670.	
	590.	510.			690.	705.	
	615.	520.			700.	720.	
	610.	570.			725.	730.	
	655.	600.			700.	740.	
	675.	600.			665.	725.	
	680.	575.			655.	795.	
	715.	570.		40 A	675.	905.	MADA-
	735.	580.			715.	915.	GASCAR
	730.	450.			840.	855.	
	660.	460.			835.	825.	
	610.	465.			800.	820.	
	590.	455.			745.	835.	
36 A	730.	450.	NAMIBIA		725.	845.	
	735.	580.			675.	905.	
	740.	605.		41 A	740.	605.	BOTSWANA
	750.	560.			790.	655.	
	820.	545.			835.	605.	
	865.	540.			825.	585.	
	865.	495.			845.	560.	
	850.	485.			820.	545.	
	795.	480.			750.	560.	
	730.	450.			740.	605.	
37 A	380.	335.	TOGO	42 A	845.	560.	REPUBLIC
	445.	340.			825.	585.	OF
	445.	325.			835.	605.	SOUTH
	385.	315.			790.	655.	AFRICA
	380.	335.			795.	680.	
38 A	710.	670.	ZIMBABWE		830.	685.	
	730.	705.			830.	695.	

850.	695.
920.	625.
940.	540.
920.	525.
865.	495.
865.	540.
820.	545.
845.	560.

99999
E-VALUES X

90.
90.
90.
78.
90.
93.
−1.
131.
−1.
153.
176.
−1.
105.
175.
146.
94.
115.
141.
143.
124.
103.
−1.
−1.
−1.
112.
152.
150.
108.
76.
103.
122.

```
                119.
                111.
                 84.
                157.
                143.
                 -1.
                 -1.
                 76.
                147.
                 63.
                 -1.
                 72.
99999
F-MAP      X
C      Infant Mortality for the Mid 1980s
C      Infant deaths per 1000 live births
C
  1 10.    10.
  2       -100.   -100.   -1100.   1100.
  6
 20  0.
 23
99999
E-VALUES X
                610.
                2530.
                1220.
                7500.
                680.
                 -1.
                370.
                 -1.
                320.
                370.
                 -1.
                620.
                140.
                390.
                270.
                200.
                200.
```

760.
810.
−1.
−1.
−1.
330.
110.
270.
150.
290.
170.
290.
240.
270.
400.
170.
200.
−1.
−1.
650.
150.
250.
−1.
2010.
99999
F-MAP   X
12
99999
999999

APPENDIX H

Input Data for GIMMS Example, Chapter 4

*FILEIN SEGMENT FILEOUT = 10 FILENAME = AFRICA
   TITLE = 'AFRICAN COUNTRIES GIMMS SEGMENTS'
   LIMITS 0,0, 1000, 1000
   BEGIN
   SEGMENTS
SEA Z01 705 860 705 825 725 765/
Z24 Z05 725 765 705 745 610 745/
Z04 Z05 610 745 600 860/
SEA Z05 600 860 645 850 705 860/
SEA Z24 725 765 770 695/
SEA Z25 770 695 830 635/
SEA Z26 830 635 850 615 930 630 925 585 880 515 810 465/
SEA Z28 810 465 785 420/
SEA Z32 785 420 795 345/
SEA Z39 795 345 800 300 725 240 730 180 695 170/
SEA Z40 905 325 915 285 855 160 825 165 820 200 835 255 845 275 905
       325/
SEA Z42 695 170 695 150 625 080 540 060 525 080 495 135/
SEA Z36 495 135 485 150 480 205 450 270/
SEA Z35 450 270 460 340 465 390 455 410/
SEA Z29 455 410 445 420/
SEA Z22 445 420 440 435/
SEA Z21 440 435 420 460 425 510/
SEA Z20 425 510 425 535/
SEA Z19 425 535 385 535 370 555 350 555/
SEA Z16 350 555 340 555/
SEA Z37 340 555 325 555/
SEA Z15 325 555 280 545/
SEA Z12 280 545 220 540/

SEA Z11 220 540 180 570/
SEA Z10 180 570 160 590/
SEA Z09 160 590 140 620/
SEA Z08 140 620 125 640/
SEA Z07 125 640 115 680 130 700/
SEA Z06 130 700 145 710 135 770/
SEA Z01 135 770 235 860 245 885 295 930 330 920/
SEA Z02 330 920 375 930 440 930/
SEA Z03 440 930 465 930 455 890 470 880/
SEA Z03 470 880 505 870 555 845 560 880 600 860/
Z26 Z25 830 635 830 600 890 585 855 540 815 535/
Z28 Z25 815 535 750 535/
Z24 Z25 750 535 720 570 750 635 750 685 770 695/
Z26 Z28 815 535 810 465/
Z28 Z27 720 530 740 500 710 470/
Z32 Z28 785 420 710 470/
Z32 Z27 710 470 690 470/
Z32 Z30 690 470 680 445/
Z31 Z30 680 445 660 445/
Z27 Z30 660 445 670 470/
Z29 Z30 670 470 690 525/
Z24 Z27 690 525 720 530/
Z24 Z28 720 530 750 535/
Z32 Z31 680 445 670 415/
Z29 Z31 670 415 660 445/
Z39 Z32 795 345 725 335/
Z34 Z32 725 335 720 365/
Z33 Z32 720 365 680 380/
Z29 Z32 680 380 670 415/
Z39 Z34 735 335 740 300 735 275 705 310/
Z33 Z34 705 310 720 365/
Z42 Z39 695 170 680 170 680 205/
Z38 Z39 680 205 705 270 670 290/
Z33 Z39 670 290 705 310/
Z42 Z38 680 205 655 210/
Z41 Z38 655 210 605 260/
Z33 Z38 605 260 670 290/
Z36 Z33 605 260 580 265/
Z35 Z33 580 265 570 285 575 320 600 325 600 345/
Z29 Z33 600 345 655 325 680 380/
Z35 Z29 600 345 570 390 520 385 510 410 455 410/
Z36 Z35 580 265 450 270/
Z41 Z36 605 260 560 250 545 180/

Z42   Z36 545 180 540 135 495 135/
Z42   Z41 655 210 605 165 585 175 560 155 545 180/
Z29   Z24 690 525 640 540/
Z23   Z24 640 540 590 610/
Z18   Z24 590 610 600 670 600 720/
Z04   Z24 600 720 610 745
Z29   723 640 540 540 525/
Z29   Z22 540 525 525 475 485 425 445 420/
Z22   Z23 540 525 510 520/
Z22   Z20 510 520 505 510 470 510/
Z21   Z20 470 510 425 510/
Z22   Z21 470 510 485 455 440 435/
Z19   Z20 425 535 430 560 490 620/
Z18   Z20 490 620 505 600 505 570/
Z23   Z20 505 570 490 540 510 520/
Z28   Z23 505 570 590 610/
Z19   Z18 490 620 480 640/
Z17   Z18 480 640 505 680 495 705/
Z04   Z18 495 750 510 760 600 720/
Z17   Z04 495 750 465 765/
Z02   Z04 465 765 440 780 440 850/
Z03   Z04 440 850 470 880/
Z03   Z02 440 850 425 885 440 930/
Z17   Z02 465 765 375 710/
Z13   Z02 375 710 280 795/
Z06   Z02 280 795 245 825/
Z01   Z02 245 825 250 850 310 860 335 880 330 920/
Z06   Z01 245 825 240 810 205 815 180 760 135 770/
Z13   Z06 280 795 265 770 265 680 180 675/
Z07   Z06 180 675 155 700 130 700/
Z13   Z07 180 675 185 640/
Z09   Z07 185 640 160 645/
Z08   Z07 160 645 125 640/
Z09   Z08 160 645 140 620/
Z13   Z09 185 640 215 640 220 610/
Z13   Z12 220 610 250 610/
Z13   Z14 250 610 275 650 330 665/
Z13   Z17 330 665 365 670 375 710/
Z09   Z10 160 600 185 610 195 595/
Z11   Z10 195 595 180 570/
Z09   Z11 195 595 220 575/
Z12   Z11 220 575 220 540/
Z11   Z12 220 575 220 610/

```
Z14  Z12 250 610 280 600/
Z15  Z12 280 600 280 545/
Z14  Z15 280 600 290 620 315 615/
Z37  Z15 315 615 328 555/
Z14  Z37 315 615 335 620/
Z16  Z37 335 620 340 555/
Z14  Z16 335 620 345 630/
Z14  Z17 345 630 330 665/
Z17  Z16 345 630 365 625/
Z19  Z16 365 625 350 555/
Z17  Z19 365 625 375 645 480 640/
END
*POLYGON FILEIN = 10 FILEOUT = 11 ALPHA EXCLUDE ZONE = SEA
*STOP
```

References

Aronoff, S. (1989): *Geographic Information Systems: A Management Perspective.* WDL Publications, Ottawa, Ontario, Canada.

Betson, R. (1976): *Urban Hydrology: A Systems Study in Knoxville, Tennessee.* Tennessee Valley Authority, Division of Water Management, Knoxville, Tennessee.

Burrough, P. A. (1986); *Principles of Geographical Information Systems for Land Resources Assessment.* Clarendon Press, Oxford.

Carruthers, A. W. (1985): *Introductory User Guide to GIMMS.* GIMMS Ltd., 30 Keir Street, Edinburgh EH3 9EU.

Chorley, R. J. and Haggett, P. (eds) (1969): *Models in Geography.* Methuen, London.

Chorley Report (1987): *Handling Geographic Information. Report to the Secretary of State for the Environment of the Committee of Enquiry into the Handling of Geographic Information.* Her Majesty's Stationery Office, London.

Cooke, D. F. (1987): Map Storage on CD ROM. *Byte Magazine,* **12**(8); July 1987, pp. 129–138.

Curran, P. J. (1985): *Principles of Remote Sensing.* Longman, London.

Date, C. J. (1986): *An Introduction to Database Systems.* Volume 1, 4th ed., Addison-Wesley, Reading, Massachusetts.

Davis, J. C. (1973): *Statistics and Data Analysis in Geology,* John Wiley and Sons, New York.

Draper, N. R. and Smith, H. (1966): *Applied Regression Analysis.* John Wiley and Sons, New York.

Ebdon, D. S. (1977): *Statistics in Geography: A Practical Approach.* Basil Blackwell, Oxford.

Edwards, A. L. (1984): *An Introduction to Linear Correlation and Regression,* 2nd ed. W. H. Freeman and Co., New York.

Evans, I. S. (1983): Univariate analysis: presenting and summarising single variables. In: Rhind, D. (ed.), *A Census User's Handbook.* Methuen, London, pp. 115–150.

Forrester, J. W. (1971): *World Dynamics* Wright-Allen Press Inc., Cambridge, Massachusetts.

Frude, N. (1987): *A Guide to SPSS/PC+.* Macmillan Education Ltd., Basingstoke.

Guttman, A. J. (1977): *Programming and Algorithms: An Introduction.* Heinemann Educational Books, London.

Harris, R. (1987): *Satellite Remote Sensing: An Introduction.* Routledge & Kegan Paul, London.

Hogg, J. and Stuart, N. (1987): Resource analysis using remote sensing and an object-orientated geographic information system. *Proc. 13th Ann. Conference, The Remote Sensing Society, Nottingham, September 1987,* pp.79–92. University of Nottingham, Remote Sensing Society.

Hollingdale, S. H. and Toothill, G. C. (1965): *Electronic Computers*: Penguin Books, Harmondsworth, Middlesex.

Holtan, H. N. and Lopez, N. C. (1970): USDAHL-70 model of watershed hydrology. *Technical Bulletin* No. 1435, Agricultural Research Service, US Department of Agriculture, US Government Printing Office, Washington, DC.

Huggett, R. (1980): *Systems Analysis in Geography*. Clarendon Press, Oxford.

Johnston, R. J. (1978): *Multivariate Statistical Analysis in Geography. A Primer on the General Linear Model*. Longman, London.

Kingslake, R. (1986): *An Introductory Course in Computer Graphics*. Chartwell-Bratt, Bromley, Kent/Studentlitteratur, Lund, Sweden.

Kirkby, M. J., Burt, T. P., Naden, P. S. and Butcher, D. P. (1975): *Computer Simulation in Physical Geography*. John Wiley and Sons, Chichester.

Lin, Gong-Yuh (1981): Simple Markov chain model of smog probability in the south coast air basin of California. *Professional Geographer*, **33**(2), 228–236.

MacDougal, E. B. (1976): *Computer Programming for Spatial Problems*. Edward Arnold, London.

Mather, P. M. (1976): *Computational Methods of Multivariate Analysis in Physical Geography*. John Wiley and Sons, Chichester.

Mather, P. M. (1989): *Computer Processing of Remotely-Sensed Images: An Introduction*, paperback edition. John Wiley and Sons, Chichester.

McGuire, D. J. (1989): *Computers in Geography*. Longman, Harlow, Essex and John Wiley and Sons, Inc., New York.

Meadows, D. H., Meadows, D. L., Randers, J. and Behrens, W. W. III (1972): *Limits of Growth*. Earth Island Ltd., London.

Miller, R. and Reddy, F. (1987): Mapping the world in Pascal. *Byte Magazine*, **12**(14), December 1987, pp. 329–334.

Monmonier, M. S. (1982): *Computer-Assisted Cartography: Principles and Prospects*. Prentice-Hall, Englewood Cliffs, New Jersey.

Mooneyhan, D. W. (1987): An overview of applications of geographical information systems within the United Nations Environment Program. *Proc. GIS'87 Symposium*, American Society for Photogrammetry and Remote Sensing, Falls Church, Virginia, pp. 536–543.

Morrison, J. L. (1980): Computer technology and cartographic change. In: *The Computer in Contemporary Cartography*, ed. D. R. Fraser Taylor. John Wiley and Sons, Chichester, pp. 5–24.

Norcliffe, G. B. (1977): *Inferential Statistics for Geographers*. Hutchinson, London.

Norusis, M. J. (1983): *SPSS^X Introductory Statistics Guide*. McGraw-Hill, New York.

Oldfield, J. V., Williams, R. D. and Wiseman, N. E. (1987): Content-addressible memories for storing and processing recursively subdivided images and trees. *Electronics Letters*, **23**, 262–263.

Peucker, T. (1972): *Computer Cartography*. Commission on College Geography, Association of American Geographers, Washington, DC.

Robinson, A. H., Morrison, J. L. and Muehrcke, P. C. (1977): Cartography 1950–2000. *Transactions of the Institute of British Geographers*, New Series, **2**, 3–18.

Rosing, K. A. and Wood, P. A. (1971): *Character of a Conurbation: A Computer Atlas of Birmingham and the Black Country*. University of London Press, London.

Shepherd, J. and Conway, K. (1988): A database for South East England. *Economic and Social Research Council Newsletter*, **63**, Working with Geographical Information Systems, October 1988, pp. 14–17.

Snyder, J. P. (1982): Map projections used by the U.S. Geological Survey. *Geological Survey Bulletin*, **1532**, second edition, US Government Printing Office, Washington, DC.

Thomas, R. H. and Huggett, R. J. (1980): *Modelling in Geography – A Mathematical Approach*. Harper and Row, London.

Thompson, M. M. (1979): *Maps for America: Cartographic Products of the U.S. Geological Survey and Others*. US Government Printing Office, Washington, DC.

US Committee on Earth Sciences (1989): *Our Changing Planet: The FY 1990 Research Plan*. Executive Office of the President, Office of Science and Technology Policy, Washington, DC.

Walker, P. A. and Grant, I. W. (1986): Quadtree: A Fortran program to extract the quadtree structure of a raster format multicolored image. *Computers and Geosciences*, **12**, 401–410.

Wilson, A. G. (1981) *Geography and the Environment: Systems Analytical Methods*. John Wiley and Sons, Chichester.

Wrigley, N. and Bennett, R. J. (1981): *Quantitative Geography: A British View*. Routledge & Kegan Paul, London.

Wyatt, B., Briggs, D. and Mounsey, H. (1988): CORINE: An information system on the state of the environment in the European Community. In: Mounsey, H. and Tomlinson, R. (eds), *Building Databases for Global Science*, Proceedings of the First Meeting of the International Geographical Union Global Database Planning Project, 9–13 May 1988. Taylor and Francis, London, pp. 378–396.

Yoeli, P. (1982): Cartographic drawing with computers. *Computer Applications*, **8**. Department of Geography, University of Nottingham.

Index